Springer Theses

Recognizing Outstanding Ph.D. Research

For further volumes:
http://www.springer.com/series/8790

Aims and Scope

The series "Springer Theses" brings together a selection of the very best Ph.D. theses from around the world and across the physical sciences. Nominated and endorsed by two recognized specialists, each published volume has been selected for its scientific excellence and the high impact of its contents for the pertinent field of research. For greater accessibility to non-specialists, the published versions include an extended introduction, as well as a foreword by the student's supervisor explaining the special relevance of the work for the field. As a whole, the series will provide a valuable resource both for newcomers to the research fields described, and for other scientists seeking detailed background information on special questions. Finally, it provides an accredited documentation of the valuable contributions made by today's younger generation of scientists.

Theses are accepted into the series by invited nomination only and must fulfill all of the following criteria

- They must be written in good English.
- The topic should fall within the confines of Chemistry, Physics, Earth Sciences, Engineering and related interdisciplinary fields such as Materials, Nanoscience, Chemical Engineering, Complex Systems and Biophysics.
- The work reported in the thesis must represent a significant scientific advance.
- If the thesis includes previously published material, permission to reproduce this must be gained from the respective copyright holder.
- They must have been examined and passed during the 12 months prior to nomination.
- Each thesis should include a foreword by the supervisor outlining the significance of its content.
- The theses should have a clearly defined structure including an introduction accessible to scientists not expert in that particular field.

Shinichiro Seki

Magnetoelectric Response in Low-Dimensional Frustrated Spin Systems

Doctoral Thesis accepted by
The University of Tokyo, Japan

Author
Dr. Shinichiro Seki
Department of Applied Physics
The University of Tokyo
Tokyo
Japan

Supervisor
Prof. Yoshinori Tokura
Department of Applied Physics
The University of Tokyo
Tokyo
Japan

ISSN 2190-5053 ISSN 2190-5061 (electronic)
ISBN 978-4-431-54682-5 ISBN 978-4-431-54091-5 (eBook)
DOI 10.1007/978-4-431-54091-5
Springer Tokyo Heidelberg New York Dordrecht London

Printed on acid-free paper

Springer is part of Springer Science+Business Media (www.springer.com)

Supervisor's Foreword

The interplay between electricity and magnetism has long been one of the central issues in the field of condensed matter physics. For conductive materials, the electron transport process was found to be significantly affected by the underlying spin texture. The discovery of giant magnetoresistance and current-induced spin torque provides the foundation for spintronics, which is now widely employed in hard disk drives and magnetoresistive random access memory (MRAM). For insulating materials, in contrast, the emergence of magnetoelectric effects, i.e., electric (magnetic) field control of magnetic (dielectric) properties, has been expected especially in compounds characterized by both magnetic and dielectric orders (multiferroics). Because the energy dissipation by an applied electric field is negligible in insulators, this approach is believed to be more energy-efficient compared with the current-driven approach in conductive materials. However, it turns out that such multiferroic materials are rather rare, and also that the coupling between both orders is very weak in general. A recent breakthrough is the discovery of the giant magnetoelectric effect in perovskite $TbMnO_3$, where spiral spin order magnetically induces ferroelectricity and application of the magnetic field leads to the flop of the spin–spiral plane and of the electric polarization direction. While the early examples of such magnetically induced ferroelectrics are structurally rather complicated, magnetic frustration, i.e., competition between different magnetic interactions, is considered the key to realizing spiral magnetism.

In the present thesis work by Shinichiro Seki, two of the most typical frustrated spin systems, i.e., triangular lattice antiferromagnets and edge-shared chain magnets, have systematically been investigated. He has found that these systems host various types of noncollinear spin texture tied with ferroelectricity, thereby enabling versatile magnetoelectric response such as electric induction of spin-helicity reversal or magnon excitation as well as magnetic control of electric polarization direction or ferroelectric domain distribution. Given the crystallographic simplicity of target systems, the above findings are rather surprising but highlight the ubiquitous local magnetoelectric coupling in solids. The current study offers a useful guideline in the search for new multiferroic materials with unique magnetoelectric functions, and also provides an important basis for the deeper understanding of magnetoelectric phenomena in more complex systems.

Tokyo, Japan, October 2011 Prof. Yoshinori Tokura

Acknowledgments

I would like to express my sincerest gratitude to Prof. Yoshinori Tokura for his continuous guidance, discussion, suggestion, and encouragement in the course of this work.

I also express my appreciation to the following people: Prof. T. Arima (diffraction experiment, discussion), Prof. S. Iguchi (discussion, encouragement), Prof. S. Ishiwata (crystal growth, dielectric measurement), Dr. H. Ito (SHG measurement), Mr. M. Uchida (discussion), Dr. D. Okuyama (diffraction experiment), Dr. Y. Onose (discussion), Dr. S. Onoda (discussion), Dr. F. Kagawa (dielectric measurement), Dr. H. Katsura (discussion), Dr. Y. Kaneko (crystal growth), Prof. T. Kida (THz spectroscopy, discussion), Mr. S. Kumakura (THz spectroscopy), Mr. T. Kurumaji (experiment, discussion), Prof. R. Kumai (diffraction experiment), Prof. H. Sawa (diffraction experiment), Mr. Y. Shiomi (crystal growth), Prof. R. Shimano (THz spectroscopy), Dr. M. Soda (diffraction experiment), Prof. Y. Taguchi (discussion), Dr. S. Tanaka (discussion), Dr. Y. Tokunaga (dielectric measurement), Prof. T. Hanasaki (discussion), Prof. T. Hasegawa (ESR measurement), Prof. K. Hirota (diffraction experiment), Prof. N. Furukawa (discussion), Dr. H. Matsui (ESR, discussion), Dr. N. Matsuura (neutron diffraction), Prof. S. Miyasaka (discussion, encouragement), Dr. S. Miyahara (discussion), Dr. H. Murakawa (dielectric measurement), Dr. M. Mochizuki (discussion), Dr. Y. Yamasaki (dielectric measurement, diffraction experiment, discussion), and Dr. M. Rahlenbeck (dielectric measurement).

I am grateful to Prof. Kyoko Ishizaka, Prof. Yoshihiro Iwasa, Prof. Yoshinori Onose, Prof. Naoto Nagaosa, and Prof. Yukitoshi Motome for valuable comments and recommendations about this thesis. I also thank all the members of Tokura Group for their help in many ways.

Finally, I would like to express special gratitude to my family for their support and encouragement over the years.

Tokyo, October 2011 Shinichiro Seki

Contents

Chapter 1
Introduction

Multiferroics and Magnetoelectric Effect

Nowadays, magnetism and electricity in solids are utilized for every aspect of modern technology. As Maxwell's equations proposed, dynamics of magnetic field (H) and electric field (H) cannot be independent but rather tightly coupled. In solids, E induces electric polarization P, and H induces magnetization M. Materials with spontaneous P (M) are called ferroelectrics (ferromagnets), and ones characterized by both ferroelectric and ferromagnetic natures are called "multiferroics". Unfortunately, such multiferroics turn out to be quite rare, and the coupling between P and M is also very weak in general. Correspondingly, H-control of P or E-control of M, which is often called "magnetoelectric (ME) effect", still remains highly nontrivial [1–3] (Fig. 1.1). Electric control of magnetism is now one of the central issues in the field of spintronics, and realization of gigantic magnetoelectric effects in solids may find wide application such as more efficient magnetic storage or H-controled FeRAM.

Quite recently, unprecedentedly large magnetoelectric effects have been discovered in a rather unexpected class of materials known as "frustrated magnet" [4–6]. Notably, these compounds host ferroelectricity of magnetic origin; Magnetic frustration sometimes reduces the symmetry of spin structure, which is now believed to affect the symmetry of charge distribution and thus induces ferroelectricity. In this thesis, we focus on triangular lattice antiferromagnets and one-dimensional chain magnets as typical examples of frustrated spin system, and investigate their magnetoelectric responses in detail.

Linear Magnetoelectric Effects

In 1894, Pierre Curie first predicted the possibility of intrinsic ME behavior on the basis of symmetry considerations [1]. In general, the free energy (F) of the system can be described using Landau expansion:

S. Seki, *Magnetoelectric Response in Low-Dimensional Frustrated Spin Systems*,
Springer Theses, DOI: 10.1007/978-4-431-54091-5_1,

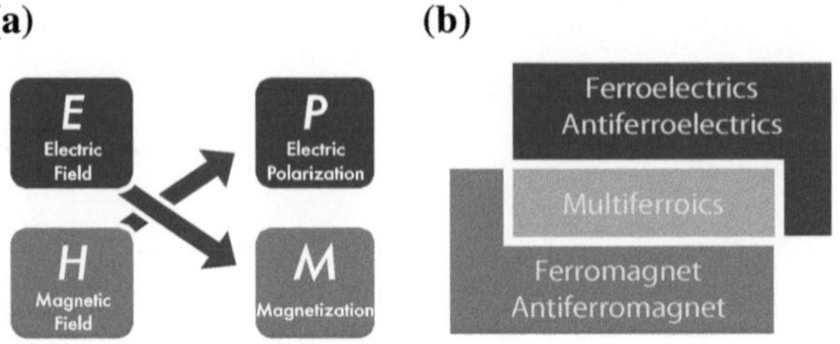

Fig. 1.1 **a** Magnetoelectric effect, and **b** multiferroics

$$-F(E,H) = \frac{1}{2}\varepsilon_{ij}E_iE_j + \frac{1}{2}\mu_{ij}H_iH_j + \alpha_{ij}E_iH_j + \frac{\beta_{ijk}}{2}E_iH_jH_k + \frac{\gamma_{ijk}}{2}H_iE_jE_k + \ldots \tag{1.1}$$

Here, ε and μ represent dielectric coefficient and magnetic permeability. α, β, and γ are material-dependent coupling tensors for corresponding orders of electric field (E) and magnetic field (H). When we assume the $E = 0$ condition, electric polarization P purely induced by H can be given as

$$P_i = -\frac{\mathrm{d}F}{\mathrm{d}E_i} = \alpha_{ij}H_j + \frac{\beta_{ijk}}{2}H_jH_k + \ldots \tag{1.2}$$

The first term suggests that P appears linearly to the applied H. Higher order magnetoelectric effect represented by the following terms can also be active, while their contribution to induced P is much smaller in case of weak H. Likewise, when we assume the $H = 0$ condition, magnetization M purely induced by E can be given as

$$M_i = -\frac{\mathrm{d}F}{\mathrm{d}H_i} = \alpha_{ji}E_j + \frac{\gamma_{ijk}}{2}E_jE_k + \ldots \tag{1.3}$$

The first term means that M appears linearly to the applied E, with the same coupling tensor α as adopted for the case of H-induced P. Here, the problem is the necessary condition to host the non-zero α. H and M change their sign for time-reversal but not for space-inversion. Contrastingly, E and P change their sign for space-inversion but not for time-reversal. Thus, non-zero α (i.e. emergence of linear magnetoelectric effect) is allowed only in the crystal where both time-reversal and space-inversion symmetry are simultaneously broken. Such linear ME effects were experimentally demonstrated in several compounds including Cr_2O_3 and $GaFeO_3$ [1, 2, 7], while it also turned out that this phenomena is quite weak in general. It was further shown that the magnitude of α is limited by the relation

$$\alpha_{ij}^2 \leq \varepsilon_{ii}\mu_{jj}, \tag{1.4}$$

which puts the cap to the enhancement of magnetoelectric response under the scheme of linear ME effects [1].

Phase Control in Multiferroics

Another promising route to achieve giant ME effects is the employment of phase control. If some specific material possesses both magnetic and ferroelectric orders and there exists strong coupling between them, H-induced magnetic phase transition (or E-induced dielectric phase transition) should significantly affect the corresponding dielectric (magnetic) properties of the same compound. Since the order parameter responds nonlinearly to the applied external field in case of the phase transition, now we don't have to care about the upper limit of linear magnetoelectric response as suggested in Eq. (1.4).

However, the successive studies clarified that such ferroelectric (anti-) ferromagnets are quite rare in nature. Magnetism requires transition metal ions with a partially filled d-shells. In contrast, stabilization of ferroelectricity via noncentrosymmetric lattice displacement requires an empty d-shell; Spins in a partially-filled d-shell are kept parallel by the Hund's rule, which breaks the strong covalent bonds between metal-3d and oxygen-2p orbitals that are necessary for ferroelectricity. Thus, magnetism and ferroelectricity are generally expected to be mutually exclusive [8, 9]. Still, there exist some exceptional cases such as ferroelectric (anti-)ferromagnets $BiFeO_3$ and $BiMnO_3$. In these compounds, however, magnetism emerges from transition metal 3d-shell and ferroelectricity does mainly from Bi 6s-shell. Since their magnetic and dielectric orders arise from different origins, these orders take place separately and almost no magnetoelectric coupling can be expected.

Magnetically-Induced Ferroelectricity and Giant ME Response

To achieve giant ME response, it is essential to find out some strong coupling mechanism between magnetism and ferroelectricity. One important breakthrough was achieved by the recent discovery of "magnetically-induced" ferroelectricity.

In 2003, Kimura et al have reported that orthorhombic $TbMnO_3$ with distorted perovskite structure shows simultaneous transition into ferroelectric and incomensurate magnetic phase [10]. Interestingly, when magnetic field is applied along the b-axis, the spontaneous electric polarization suddenly changes its direction by 90° from $P \parallel c$ to $P \parallel a$. The successive neutron diffraction study revealed that the observed ferroelectricity is coupled with the cycloidal magnetic order [11, 12], and directional change of P coincides with the 90°-flop of spin-spiral plane (Fig. 1.2) [13].

Later, similar strong coupling between ferroelectricity and spiral (or some other complex) magnetic orders has been discovered in several frustrated magnets such as $R$$MnO_3$ [14, 15], $R$$Mn_2O_5$ [16], $Ni_3V_2O_8$ [17], $MnWO_4$ [18], $CoCr_2O_4$ [19], $CuFeO_2$ [20], $LiCu_2O_2$ [21], $(Ba, Sr)_2Zn_2Fe_{12}O_{22}$ [22], and CuO [23]. As in the case of $TbMnO_3$, they often exhibit unprecedentedly large and versatile magnetoelectric

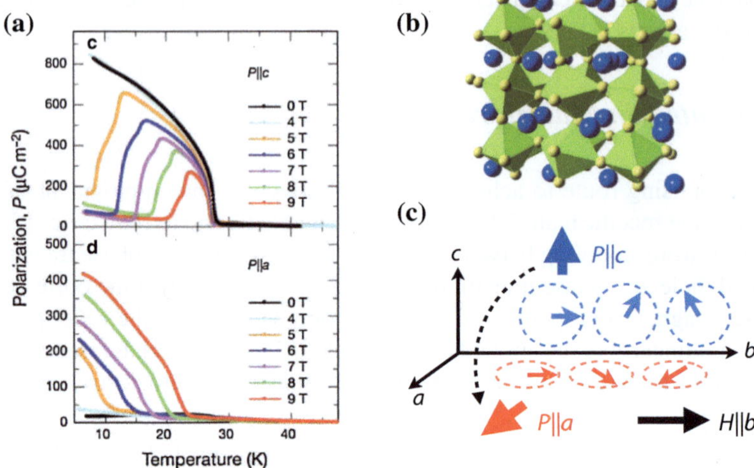

Fig. 1.2 **a** Temperature dependence of electric polarization P for $TbMnO_3$ under various strength of magnetic field applied parallel to the b-axis. **b** Distorted perovskite structure. **c** Change of magnetic structure and associated P-direction under $H \parallel b$ for $TbMnO_3$. (Adapted with permission from [10], ©2003 Nature Publishing Group.)

responses such as flop, reversal, or rotation of P under applied H. Most of these effects turned out to be associated with the change of magnetic structure.

In this thesis, we mainly focus on such ferroelectric helimagnets with strong magnetoelectric coupling. In the following, we introduce the origin of spiral magnetic orders and the coupling mechanism between ferroelectricity and magnetism.

Magnetic Frustration

In a simple localized spin system, the spin Hamiltonian $\mathcal{H}_{\rm mag}$ can be given as

$$\mathcal{H}_{\rm mag} = -\sum_{i,j} J_{ij}(\mathbf{S}_i \cdot \mathbf{S}_j) - \sum_{i,j} \mathbf{D}^{\rm DM} \cdot (\mathbf{S}_i \times \mathbf{S}_j) - D\sum_i S_{iz}^2 - g\mu_{\rm B}\sum_i \mathbf{H}\cdot\mathbf{S}_i \quad (1.5)$$

Here, $\mathbf{S}_i$ is the electron spin at site i, and S_{iz} is its z-component. The first, second, third, and forth terms represent Heisenberg exchange interaction, Dzyaloshinskii-Moriya interaction, single-ion anisotropy, and Zeeman energy under applied magnetic field **H**, respectively. Among them, only the second and third terms arise from the spin-orbit interaction.

One important source of spiral magnetic order is Dzyaloshinskii-Moriya interaction, which is allowed to exist only when space-inversion symmetry is broken at the center of two adjacent magnetic sites [24]. This interaction favors orthogonal spin

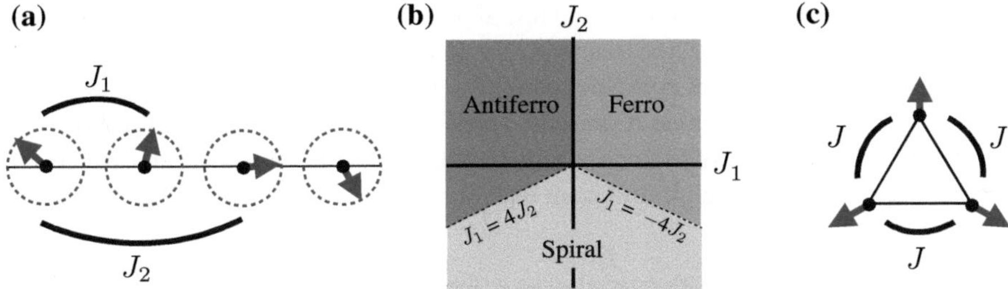

Fig. 1.3 **a** One dimensional chain with competing J_1 and J_2. **b** The magnetic ground state for J_1 vs J_2 model. **c** Triangular lattice antiferromagnet with 120°-spin order

arrangement within a plane perpendicular to the $\mathbf{D}^{\mathrm{DM}}$-vector, and is considered as the origin of helimagnetism in several non-centrosymmetric magnets.

On the other hand, Heisenberg exchange interaction can also stabilize spiral magnetic order even without contribution of Dzyaloshinskii-Moriya interaction. While Heisenberg exchange interaction generally favors parallel or antiparallel spin arrangement depending on the sign of J_{ij}, in some lattices it is not possible to satisfy this condition for all bonds. Such a competition is called "magnetic frustration", and often leads to complex magnetic order to minimize the total energy. Notably, all magnetically-induced ferroelectrics have been reported to contain magnetic frustration as a source of helimagnetic or other long-period magnetic orders. In the following, we show some simple examples of frustrated spin systems.

1D-Chain with Competing Nearest and Next-Nearest Neighbor Interactions

First, we assume one-dimensional chain magnet with nearest neighbor interaction J_1 and next-nearest neighbor interaction J_2 (Fig. 1.3a). For simplicity, we only consider Heisenberg exchange interactions. The spin Hamiltonian $\mathcal{H}_{\mathrm{mag}}$ can be described as

$$\mathcal{H}_{\mathrm{mag}} = -J_1 \sum_{\mathrm{N.N.}} (\mathbf{S}_i \cdot \mathbf{S}_j) - J_2 \sum_{\mathrm{N.N.N.}} (\mathbf{S}_i \cdot \mathbf{S}_k) \tag{1.6}$$

When J_2 is negative (i.e. antiferromagnetic), neither ferromagnetic nor simple antiferromagnetic spin arrangement can satisfy the second term and thus magnetic frustration arises. Here, we introduce a generalized spin order

$$\mathbf{S}_i = \begin{pmatrix} S_{ix} \\ S_{iy} \\ S_{iz} \end{pmatrix} = \begin{pmatrix} S \cdot \cos(qr_i + \varphi) \\ S \cdot \sin(qr_i + \varphi) \\ 0 \end{pmatrix}, \tag{1.7}$$

where q and φ are the wave number and initial phase of magnetic modulation, and r_i is the position of ith magnetic site, respectively. If we define $r = r_{i+1} - r_i$ and $\theta = rq$, ferromagnetic arrangement corresponds to $\theta = 0$ and simple antiferromagnetic one does to $\theta = \pi$. Other θ-value represents spiral spin order. By substituing Eq. (1.7) into Eq. (1.6) and analyzing the condition to satisfy $d\mathcal{H}/d\theta = 0$, we can deduce the wave number of magnetic ground state (θ_0).

$$\cos\theta_0 = -\frac{J_1}{4J_2} \qquad (-1 \le \cos\theta_0 \le 1) \tag{1.8}$$

We can see that spiral magnetic order is more stable than ferromagnetic or simple antiferromagnetic ones, as long as the $(-1 \le \cos\theta_0 \le 1)$ condition is satisfied. The period of spin-spiral depends on the J_1/J_2 ratio. When we cannot define θ_0 as real number, ferromagnetic or simple antiferromagnetic spin arrangement becomes the ground state. These results are summarized in a $J_1 - J_2$ phase diagram (Fig. 1.3b) [25].

Geometrical Frustration

Even when we consider only the nearest neighbor interaction J, magnetic frustration can take place depending on the geometry of underlying lattice (Geometrical frustration). One of typical examples is the triangular lattice antiferromagnet. In this case, anti-parallel spin arrangement cannot be realized simultaneously for all three bonds. Instead, 120°-spin order, in which each spin is aligned at 120° to its neighboring spins, is reported to become the ground state (Fig. 1.3c) [26]. Similar geometrical frustration can also be realized in some other lattices, such as Kagomé, pyrochlore, and f.c.c [27].

Role of Magnetic Anisotropy

In the above two examples, we considered the competition of Heisenberg exchange interactions. They only determine the relative angle between neighboring magnetic moments, and never specify the favorable spin-direction. In general, the orientation of spin-spiral plane is determined by the magnetic anisotropy, which strongly depends on the electronic state and crystal field at the magnetic site. In Fig. 1.4, several variations of helimagnetic orders are indicated. Cycloidal and proper screw magnetic orders corresponds to the one without net magnetization, with spin-spiral plane parallel and perpendicular to the magnetic modulation vector (q), respectively. The one with net magnetization is called conical spin order. By applying magnetic field, it is often possible to reorient the spin-spiral plane against the magnetic anisotropy

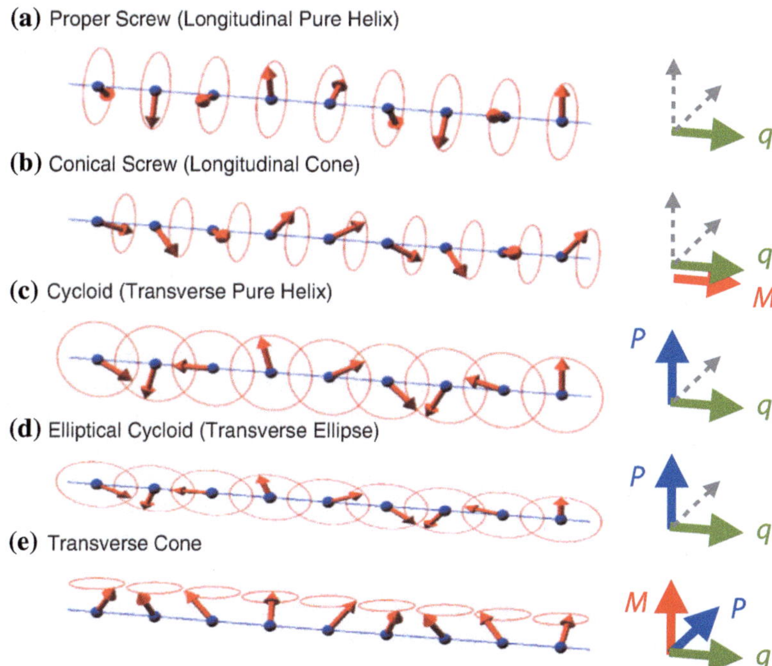

Fig. 1.4 Various types of spiral spin orders. M, q, and P denote the direction of magnetization, magnetic modulation vector, and electric polarization expected from the inverse D-M model as given by Eq. (1.10)

(Spin-flop transition). If magnetic anisotropy is sufficiently strong, collinear magnetic order like ↑↑↓↓ becomes more favorable than simple helimagnetic one.

Magnetically-Induced Ferroelectricity

Magnetic Symmetry

To induce ferroelectricity, the symmetry of charge distribution must be polar. While this requires the violation of space-inversion symmetry, most of magnetically-induced ferroelectrics are centrosymmetric in their paramagnetic phases. Interestingly, magnetic frustration often stabilizes the spin structure with symmetry lower than that of underlying crystal lattices. When there exists some coupling mechanism to connect the spin and charge degrees of freedom, such a symmetry reduction is reflected in the charge distribution and thus leads to ferroelectricity.

In case of $TbMnO_3$, $P \parallel c$ is induced by bc-cycloid spin order modulating along the b-axis (Fig. 1.2c) [11]. Hereafter, we define the spiral axis as the direction

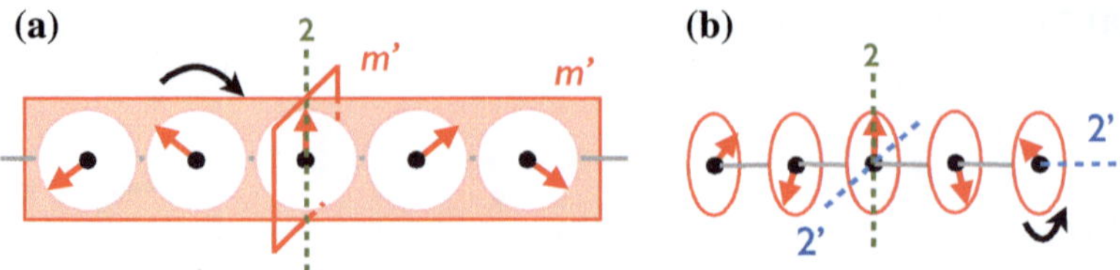

Fig. 1.5 **a** Cycloidal spin structure and **b** proper screw spin structure. The symmetry elements for these spin configurations are also indicated

perpendicular to the spin-spiral plane. Figure 1.5a indicates the symmetry operations compatible with the cycloidal spin order. It possesses a m'-plane (time-reversal followed by a mirror reflection) normal to the modulation vector, two-fold rotation axis (2) perpendicular both to the modulation vector and to the spiral axis, and m'-plane normal to the spiral axis. No other symmetry operation is allowed. Thus, cycloidal magnetic structure always has polar axis along the 2-axis and ferroelectricity can appear along this direction. This symmetry analysis well reproduces the P-direction as experimentally observed for $TbMnO_3$.

In contrast, proper screw magnetic order possesses $2'$-axis (time-reversal followed by two-fold rotation) parallel to the modulation vector, and 2- and $2'$-axes perpendicular to the magnetic modulation vector (Fig. 1.5b). In this case, the spin structure is non-polar and ferroelectricity cannot be induced in general.

Note that both types of spiral spin orders break space-inversion symmetry. When the space-inversion operation is applied to these spin spirals, the manner of spin-ration is reversed. This characteristic degree of freedom distinguishing the clockwise and counter-clockwise manner is called "chirality", and often plays an important role to understand the behavior of spiral magnets.

Microscopic Origin of Magnetoelectric Coupling

Since the symmetry analysis tells us nothing about the magnitude of induced polarization, we have to identify the microscopic origin of magnetoelectric coupling to guarantee the non-zero P-value. So far, at least two microscopic origins of magnetoelectric coupling have been established; exchange striction and spin-orbit coupling.

Exchange striction originates from the Heisenberg exchange interaction, and induces local polarization $\mathbf{P}_{ij}$ between two magnetic sites ($\mathbf{S}_i$ and $\mathbf{S}_j$) in form of

$$\mathbf{P}_{ij} = \boldsymbol{\pi}_{ij}(\mathbf{S}_i \cdot \mathbf{S}_j). \tag{1.9}$$

Here, $\boldsymbol{\pi}_{ij}$ essentially depends on the geometry of underlying lattice. As a simple example, we assume the situation where two different magnetic ions are alternatively aligned and $\uparrow\uparrow\downarrow\downarrow$ collinear magnetic order is realized on them (Fig. 1.6a). While the original lattice structure itself is centrosymmetric, the exchange striction

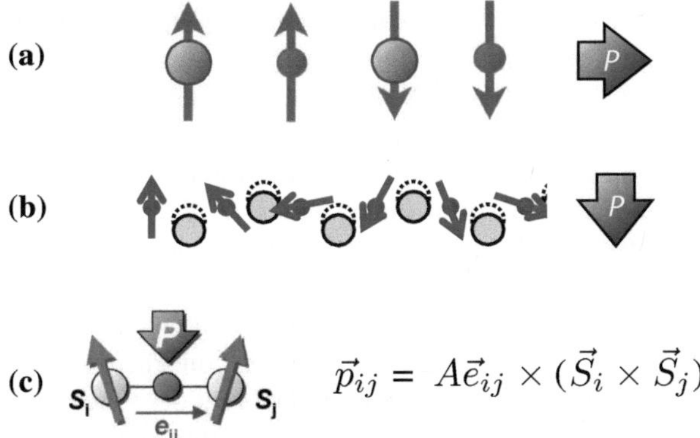

Fig. 1.6 **a** ↑↑↓↓ spin order on ABAB... lattice. Electric polarization is induced by the exchange striction. **b** The schematic illustration of the inverse D-M mechanism. **c** The local configurations assumed for Eq. (1.10)

shifts each magnetic ion in a way that optimize the energy gain by the antiferromagnetic Heisenberg interaction; ions with antiparallel spins are pulled to each other, whereas ions having parallel spins move away from each other. This lattice distortion patterns breaks space-inversion symmetry and induces net polarization along the chain direction. Such a situation is actually realized in Ca_3CoMnO_6 [28], and the present model well reproduces the experimentally reported dielectric properties. Exchange striction mechanism is also adopted to explain the magnetically-induced ferroelectricity in several canted-antiferromagnets such as $GdFeO_3$ [29] and $DyFeO_3$ [30].

Another important source to connect magnetism and charge distribution is the spin-orbit coupling. As already mentioned, spin-orbit coupling is the primal origin of Dzyaloshinskii-Moriya interaction ($\mathcal{H} = \mathbf{D}^{\mathrm{DM}} \cdot (\mathbf{S}_i \times \mathbf{S}_j)$) and magnetic anisotropy. When we assume the cycloidal spin order, $(\mathbf{S}_i \times \mathbf{S}_j)$ becomes constant for all neighboring bonds. This induces the uniform displacement of ions and/or electron clouds to maximize the energy gain by Dzyaloshinskii-Moriya interaction, which eventually leads to the induction of ferroelectricity (Fig. 1.6b). This scheme is called inverse Dzyaloshinskii-Moriya (D-M) mechanism, and the successive theoretical studies [31–33] have suggested that the local polarization $\mathbf{P}_{ij}$ produced between two magnetic sites is generally given as

$$\mathbf{P}_{ij} = A\mathbf{e}_{ij} \times (\mathbf{S}_i \times \mathbf{S}_j). \tag{1.10}$$

Here, $\mathbf{e}_{ij}$ is an unit vector connecting two magnetic sites, and A is a coupling coefficient related to the spin-orbit interaction (Fig. 1.6c). $(\mathbf{S}_i \times \mathbf{S}_j)$ is a vector perpendicular to the spin-spiral plane, and often called vector spin chirality because its sign corresponds to the clock-wise or counter-clockwise manner of spin rotation. This model predicts that cycloidal spin order induces polarization along the direction normal both to the magnetic modulation vector and to the spiral-axis (Fig. 1.4c). In case of proper screw spin order, no ferroelectric polarization can be expected due to the relationship ($\mathbf{q}$ $\parallel$)$\mathbf{e}_{ij}$ $\parallel$ ($\mathbf{S}_i \times \mathbf{S}_j$) (Fig. 1.4 a). These predictions are consistent with the suggestions of symmetry analysis. According to Eq. (1.10), H-induced reorientation of spin-spiral plane leads to directional change of P. 90°-flop of P-vector and associated spin spiral plane under $H \parallel b$ as reported in $TbMnO_3$ (Fig. 1.2c) can be well reproduced by the present scheme. Equation (1.10) also suggests that the sign of $\mathbf{P}_{ij}$ and ($\mathbf{S}_i \times \mathbf{S}_j$) is coupled; this means the reversal of P-direction always gives opposite manner of spin rotation. Such coupling between spin chirality and the sign of P has experimentally been confirmed by polarized neutron scattering study on $TbMnO_3$ [34]. The inverse D-M model successfully explains the ferroelectric and magnetoelectric behavior in many other spiral magnets (Table 1.1), and spin-orbit interaction is now believed to be the prime source of magnetoelectric coupling for helimagnets.[1] Note that there still exist several ferroelectric helimagnets whose ME properties cannot be explained within the framework of inverse D-M mechanism. For these compounds, some alternative ME coupling mechanism originating from the spin-orbit interaction has been suggested, as later mentioned in "Introduction".

Purpose

As discussed in the last section, magnetic frustration sometimes reduces the symmetry of spin structure, which is now believed to affect the symmetry of charge distribution and thus induce ferroelectricity. Here, the important is the microscopic origin of such a strong magnetoelectric coupling. Unfortunately, early examples of magnetically-induced ferroelectrics are structurally rather complicated, which often prevents the elucidation of microscopic ME coupling mechanism or generalization of observed magnetoelectric behaviors. As discussed in "Magnetic Frustration", triangular lattice with antiferromagnetic interaction and one-dimensional chain with competing J_1 and J_2 are known as the simplest examples of frustrated spin system. In this thesis, we focus on the two typical structural units to realize such a situation, and investigate their magnetoelectric response in detail.

In Chap. 3, we study the antiferromagnets with the stacked triangular lattices as shown in Fig. 1.7a. Interestingly, the geometry of triangular lattice often allows the appearance of magnetically-induced ferroelectricity that cannot be explained by either exchange striction or inverse D-M mechanism. We examined the magnetoelectric response of various types of spin orders on triangular lattices, and also performed several experiments from the viewpoints of impurity-doping effect, domain control, and dynamics.

[1] Recent ab initio calculation for $TbMnO_3$ based on the density-functional theory also suggested that non-zero P-value is obtained only when spin-orbit interaction is introduced [35, 36].

Table 1.1 Partial list of reported ferroelectric helimagnets. For $ZnCr_2Se_4$, finite P is induced only when spin-spiral plane is tilted from original proper screw configuration by applied H

	Structure	Spin order	S	T_{FE}	P ($\mu C/m^2$)	Inverse D-M	Ref.
$RMnO_3$ (R = Tb, Dy)	Orthorhombic	Cycloid	2	~28 K	~2000	○	[10, 14, 15]
$Ni_3V_2O_8$	Orthorhombic	Cycloid	1	6 K	100	○	[17]
$CoCr_2O_4$	Cubic	Conical	3/2	26 K	2	○	[19]
$MnWO_4$	Orthorhombic	Cycloid	5/2	12 K	150	○	[18]
$LiCu_2O_2$	Orthorhombic	Cycloid	1/2	23 K	8	?	[21]
$LiCuVO_4$	Orthorhombic	Cycloid	1/2	2 K	30	○?	[37]
$ZnCr_2Se_4$	Cubic	Proper screw	3/2	20 K	20	○	[38]
CuO	Monoclinic	Cycloid	1/2	230 K	150	○?	[23]
$(Ba, Sr)_2 Zn_2Fe_{12}O_{22}$	Trigonal	Proper screw	5/2	~R.T.	150	?	[22]
$Ba_2Mg_2 Fe_{12}O_{22}$	Trigonal	Conical	5/2	195 K	80	○	[39]
$CuFeO_2$	Trigonal	Proper screw	5/2	11 K	300	×	[20]
$RbFe(MoO_4)_2$	Trigonal	120°	5/2	3.8 K	5	×	[40]

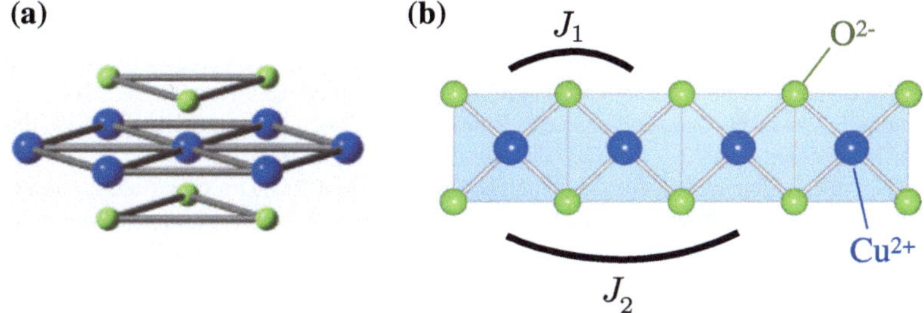

Fig. 1.7 **a** Stacked triangular lattice and **b** edge-shared CuO_2 chain

In Chap. 4, we study the helimagnets with the edge-shared CuO_2 chain structures as shown in Fig. 1.7b. While they are seemingly ideal model system to testify the validity of the inverse D-M mechanism, the early two examples of such chain helimagnets ($LiCu_2O_2$ and $LiCuVO_4$) are reported to show magnetoelectric response inconsistent with the prediction of the inverse D-M model. Since their magnetism is dominated by Cu^{2+} with $S = 1/2$, such anomalous behavior may possibly be caused by quantum fluctuation of electron spin. To clarify the origin of magnetoelectric coupling in $S = 1/2$ chain helimagnets, we examined the relationship between spin structure and induced P-directions on several related compounds through the polarized neutron scattering and detailed dielectric measurements.

In Chap. 5, we summarize the obtained results and conclude this thesis.

References

1. Fiebig M (2005) Revival of the magnetoelectric effect. J Phys D Appl Phys 38:R123
2. Smolenskii GA, Chupis IE (1982) Ferroelectromagnets. Sov Phys Usp 25:475
3. Schmid H (2008) Some symmetry aspects of ferroics and single phase multiferroics. J Phys Condens Matter 20:434201
4. Tokura Y (2006) Multiferroics as quantum electromagnets. Science 312:1481
5. Cheong S-W, Mostovoy M (2007) Multiferroics: a magnetic twist for ferroelectricity. Nat Mater 6:13
6. Eerenstein W, Mathur ND, Scott JF (2006) Multiferroic and magnetoelectric materials. Nat Mater 442:759
7. Astrov DN (1960) The magnetoelectric effect in antiferromagnetics. Sov Phys JETP 11:708
8. Hill NA (2000) Why are there so few magnetic ferroelectrics? J Phys Chem B 104:6694
9. Khomskii DI (2001) Magnetism and ferroelectricity: why do they so seldom coexist? Bull Am Phys Soc C21:002
10. Kimura T, Goto T, Shintani H, Ishizaka K, Arima T, Tokura Y (2003) Magnetic control of ferroelectric polarization. Nature (London) 426:55
11. Kenzelmann M, Harris AB, Jonas S, Broholm C, Schefer J, Kim SB, Zhang CL, Cheong S-W, Vajk OP, Lynn JW (2005) Magnetic inversion symmetry breaking and ferroelectricity in $TbMnO_3$. Phys Rev Lett 95:087206
12. Arima T, Tokunaga A, Goto T, Kimura H, Noda Y, Tokura Y (2006) Collinear to spiral spin transformation without changing the modulation wavelength upon ferroelectric transition in $Tb_{1-x}Dy_xMnO_3$. Phys Rev Lett 96:097202
13. Aliouane N, Schmalzl K, Senff D, Maljuk A, Prokes K, Braden M, Argyriou DN (2009) Flop of electric polarization driven by the flop of the Mn spin Cycloid in multiferroic $TbMnO_3$. Phys Rev Lett 102:207205
14. Goto T, Kimura T, Lawes G, Ramirez AP, Tokura Y (2004) Ferroelectricity and giant magnetocapacitance in perovskite Rare-Earth manganites. Phys Rev Lett 92:257201
15. Kimura T, Lawes G, Goto T, Tokura Y, Ramirez AP (2005) Magnetoelectric phase diagrams of orthorhombic $R\mathrm{MnO}_3$ (R = Gd, Tb, and Dy). Phys Rev B 71:224425
16. Hur N, Park S, Sharma PA, Ahn JS, Guha S, Cheong S-W (2004) Electric polarization reversal and memory in a multiferroic material induced by magnetic fields. Nature (London) 429:392
17. Lawes G, Harris AB, Kimura T, Rogado N, Cava RJ, Aharony A, Entin-Wohlman O, Yildrim T, Kenzelmann M, Broholm C, Ramirez AP (2005) Magnetically driven ferroelectric order in $Ni_3V_2O_8$. Phys Rev Lett 95:087205
18. Taniguchi K, Abe N, Takenobu T, Iwasa Y, Arima T (2006) Ferroelectric polarization flop in a frustrated magnet $MnWO_4$ induced by a magnetic field. Phys Rev Lett 97:097203
19. Yamasaki Y, Miyasaka S, Kaneko Y, He J-P, Arima T, Tokura Y (2006) Magnetic reversal of the ferroelectric polarization in a multiferroic spinel oxide. Phys Rev Lett 96:207204
20. Kimura T, Lashley JC, Ramirez AP (2006) Inversion-symmetry breaking in the noncollinear magnetic phase of the triangular-lattice antiferromagnet $CuFeO_2$. Phys Rev B 73:220401(R)
21. Park S, Choi YJ, Zhang CL, Cheong S-W (2007) Ferroelectricity in an $S = 1/2$ chain cuprate. Phys Rev Lett 98:057601
22. Kimura T, Lawes G, Ramirez AP (2005) Electric polarization rotation in a hexaferrite with long-wavelength magnetic structures. Phys Rev Lett 94:137201
23. Kimura T, Sekio Y, Nakamura H, Siegrist T, Ramirez AP (2008) Cupric oxide as an induced-multiferroic with high-T_C. Nat Mater 7:291
24. Moriya T (1960) Anisotropic superexchange interaction and weak ferromagnetism. Phys Rev 120:91
25. Nagamiya T (1968) Helical spin ordering-1 theory of helical spin configurations. Solid state phys 20:305
26. Miyashita S, Shiba H (1984) Nature of the phase transition of the two-dimensional antiferromagnetic plane rotator model on the triangular lattice. J Phys Soc Jpn 53:1145

27. Greedan JE (2001) Geometrically frustrated magnetic materials. J Mater Chem 11:37
28. Choi YJ, Yi HT, Lee S, Huang Q, Kiryukhin V, Cheong S-W (2008) Ferroelectricity in an Ising chain magnet. Phys Rev Lett 100:047601
29. Tokunaga Y, Furukawa N, Sakai H, Taguchi Y, Arima T, Tokura Y (2009) Composite domain walls in a multiferroic perovskite ferrite. Nat Mater 8:558
30. Tokunaga Y, Iguchi S, Arima T, Tokura Y (2008) Magnetic-field-induced ferroelectric state in $DyFeO_3$. Phys Rev Lett 101:097205
31. Katsura H, Nagaosa N, Balatsky AV (2005) Spin current and magnetoelectric effect in noncollinear magnets. Phys Rev Lett 95:057205
32. Sergienko IA, Dagotto E (2006) Role of the Dzyaloshinskii-Moriya interaction in multiferroic perovskites. Phys Rev B 73:094434
33. Mostovoy M (2006) Ferroelectricity in spiral magnets. Phys Rev Lett 96:067601
34. Yamasaki Y, Sagayama H, Goto T, Matsuura M, Hirota K, Arima T, Tokura Y (2007) Electric control of spin helicity in a magnetic ferroelectric. Phys Rev Lett 98:147204
35. Xiang HJ, Wei SH, Whangbo M-H, Da Silva JLF (2008) Spin-orbit coupling and ion displacements in multiferroic $TbMnO_3$. Phys Rev Lett 101:037209
36. Malashevich A, Vanderbilt D (2008) First principles study of improper ferroelectricity in $TbMnO_3$. Phys Rev Lett 101:037210
37. Naito Y, Sato K, Yasui Y, Kobayashi Y, Kobayashi Y, Sato M (2007) Ferroelectric transition induced by the incommensurate magnetic ordering in $LiCuVO_4$ J Phys Soc Jpn 76:023708
38. Murakawa H, Onose Y, Ohgushi K, Ishiwata S, Tokura Y (2008) Generation of electric polarization with rotating magnetic field in helimagnet $ZnCr_2Se_4$ J Phys Soc Jpn 77:043709
39. Ishiwata S, Taguchi Y, Murakawa H, Onose Y, Tokura Y (2008) Low-magnetic-field control of electric polarization vector in a helimagnet. Science 319:1643
40. Kenzelmann M, Lawes G, Harris AB, Gasparovic G, Broholm C, Ramirez AP, Jorge GA, Jaime M, Park S, Huang Q, Shapiro AY, Demianets LA (2007) Direct transition from a disordered to a multiferroic phase on a triangular lattice. Phys Rev Lett 98:267205

Chapter 2
Experimental Methods

Crystal Growth

All the oxide compounds listed below were identified with power X-ray diffraction. The orientation of single crystals was confirmed by Laue X-ray photography. Depending on the purpose, they were cut into rectangular shapes using a crystal cutter (for oxides) or razor's edge (for halides).

$CuFeO_2$, $CuFe_{1-x}Al_xO_2$ and $CuFe_{1-x}Ga_xO_2$

Single crystals of $CuFeO_2$, $CuFe_{1-x}Al_xO_2$ ($x = 0.01, 0.02$), and $CuFe_{1-x}Ga_xO_2$ ($x = 0.01, 0.035$) were grown by a floating zone method. As starting materials, powders of Cu_2O, Fe_2O_3, Al_2O_3, and Ga_2O_3 were used. Stoichiometric amounts of powders were mixed, ground, and sintered at 950 °C for 24 h in Ar atmosphere. The obtained polycrystal was pressed into a rod-shape. The single crystal growth was carried out in an infrared radiation furnace, with Ar atmosphere and an upper (lower) zone speed of 1.5 mm/h (3.0 mm/h), respectively. Since $CuFeO_2$ partially decomposes above 1, 180 °C, the first several cm of the crystal includes excess Fe_2O_3. As the growth goes on, composition of the molten zone gradually changes and finally the pure $CuFeO_2$ (or $CuFe_{1-x}Al_xO_2$ / $CuFe_{1-x}Ga_xO_2$) phase becomes stable [1]. Thus, we picked up only the last 5 cm of the obtained crystal. Because the doping of Al or Ga seems to slow down the convergence of crystallographic domains, we checked the single domain nature of each specimen under polarized optical microscope.[1]

[1] To observe the crystallographic domain structure under the polarized optical microscope, we require a flat mirror surface. For this purpose, the specimen of $CuFeO_2$ was polished using Al_2O_3 powder with radius down to 0.3 μm

S. Seki, *Magnetoelectric Response in Low-Dimensional Frustrated Spin Systems*, Springer Theses, DOI: 10.1007/978-4-431-54091-5_2, © Springer Japan 2012

$ACrO_2$ (A = Cu, Ag, Li and Na)

Single crystals of $CuCrO_2$ were grown by a Bi_2O_3-flux method. Powders of CuO (4.1 mg) and Cr_2O_3 (3.9 mg) were put into a Pt crucible (50 cc), and then Bi_2O_3-powder was added to fill up about 4/5 of the crucible. To melt the mixture of powders, they were heated in air atmosphere at 1, 280 °C for 24 h. The crucible was cooled slowly (1 °C/h) down to 750 °C, and then done rapidly (150 °C/h) down to the room temperature. By this method, we can obtain plate-shaped $CuCrO_2$ single crystals with widest faces parallel to the (001)-plane. The maximum size of the crystal was $7 \times 4 \times 1.5$ mm.

Polycrystals of $AgCrO_2$, $LiCrO_2$, and $NaCrO_2$ were prepared by solid state reaction from the stoichiometric mixture of Ag, Li_2CO_3, Na_2CO_3 and Cr_2O_3. They were heated at 900 °C for 48 h in O_2, at 1, 200 °C for 24 h in air, and at 1, 100 °C for 30 h in Ar, respectively. Powder x-ray diffraction measurements showed no detectable impurity, except a trace of Ag in $AgCrO_2$ specimen and of Cr_2O_3 in $NaCrO_2$ specimen. They were pressed into rods, sintered with additional heating, and cut into thin plates.

$LiCu_2O_2$ and $NaCu_2O_2$

Single crystals of $LiCu_2O_2$ were grown by a self-flux method. Stoichiometric amounts of Li_2CO_3 and CuO were put into an alumina crucible, and heated up to 1, 100 °C in air atmosphere. They were cooled slowly (2.5 °C/h) down to 930 °C, and then done rapidly (900 °C/h) down to the room temperature. Obtained $LiCu_2O_2$ single crystal is plate-shaped, with widest faces parallel to the (001)-plane. They also frequently possess additional cleavage planes parallel to (210). Due to the $a \sim 2b$ relationship of lattice constants, the fine twin structure with mixing of the a and b-axis domains was observed under a polarized optical microscope [2]. All the measurements on $LiCu_2O_2$ in this thesis were performed for the ab-twinned specimen.[2]

Single crystals of $NaCu_2O_2$, which were grown by a self-flux method [3], were provided by Prof. Keimer's group at Max Planck Institute. Since $NaCu_2O_2$ is relatively sensitive to humidity, the crystal was stored in an evacuated desiccator.

$CuCl_2$, VCl_2, MnI_2 and CoI_2

Single crystals of $CuCl_2$, MnI_2, and CoI_2 were grown by a Bridgeman method. Powder of each compound was sealed into an evacuated quartz tube ($\phi = 15$ mm) with V-shaped end. They were slowly dragged down (2.5 mm/h) through the

[2] With inappropriate conditions, unintensional growth of $LiCu_3O_3$ and/or Li_2CuO_2 was often confirmed. Since they don't have the twin structure as found in $LiCu_2O_2$, unintended phases can be easily distinguished under polarized optical microscope.

temperature gradient (10 °C/cm) produced by two-zone furnace. The temperature of upper (hotter) heater was set 700 °C for $CuCl_2$, 550 °C for CoI_2, and 700 °C for MnI_2, respectively.

Single crystals of VCl_2 were purchased from Mitsuwa Chemical Co., Ltd. They possess flake-like shape with thickness of ~0.3 mm and widest faces parallel to the (001)-plane, probably grown by the chemical vapor transport method.

Since most of MX_2-type halides are extremely sensitive to the humidity, all the handling of the specimen were performed in an Ar-filled glove box. When we measure these compounds outside of the glove box, whole surface of the specimen was covered by small amount of Apiezon-N grease to shut out the moisture in air.

Magnetic Property Measurement

Magnetization measurement was performed using a SQUID magnetometer.

Dielectric Property Measurement

For dielectric property measurements, silver paste was painted on the parallel surfaces of the specimen as electrodes. We often employed two sets of electrodes to simultaneously measure two orthogonal components of electric polarization vector.

Dielectric Constant

Dielectric constant is deduced using a LCR-meter (Agilent E4980A) in the frequency range of 1 MHz $\sim$ 1 kHz. In general, the relationship among dielectric constant ($\varepsilon^{\omega} = \varepsilon_1 + i\varepsilon_2$), electric conductivity ($\sigma^{\omega} = \sigma_1 + i\sigma_2$), and admittance ($Y^{\omega}$) is given by the following equations.

$$j^{\omega} = \sigma^{\omega} \cdot E^{\omega} \tag{2.1}$$

$$\sigma^{\omega} = i\omega\varepsilon^{\omega} \tag{2.2}$$

$$Y^{\omega} = \sigma^{\omega} \cdot \frac{S}{l} \tag{2.3}$$

Here, j^{ω}, E^{ω}, S, and l represent current density, electric field, area of electrode, and thickness of the specimen, respectively. When the specimen can be considered as a RC-circuit with parallel connection of resistivity R_p and capacitance C_p, its admittance Y^{ω} can be given as $Y^{\omega} = (1/R_p) + i\omega C_p$. LCR-meter measures Y^{ω}, and displays the values of C_p and $D = 1/(2\pi\omega C_p R_p)$. As a result, ε_1 and ε_2 can be obtained by the following equations.

$$\varepsilon_1 = \frac{C_p l}{S} \tag{2.4}$$

$$\varepsilon_2 = \frac{2\pi l C_p D}{S} \tag{2.5}$$

Electric Polarization

Electric polarization (P) is deduced using an electrometer (Keithley Model 6517A). When the magnitude of P changes as a function of time (t), the rearrangement of surface charge induces polarization current (J).

$$\frac{\mathrm{d}P}{\mathrm{d}t} = j = \frac{J}{S} \tag{2.6}$$

Here, j and S indicate polarization current density and area of electrode. Electrometer can measure the polarization current in an accuracy of sub-pA. To obtain the magnitude of P, we measured J with constant rates of temperature-sweep (2 $\sim$ 20 K/min), H-sweep (50 $\sim$ 131 Oe/sec), or H-rotation (0.5° $\sim$ 2°/sec), and integrated it with time.

To enlarge the population of specific P-domains, the poling electric field (50 $\sim$ 300 kV/m) was applied along the appropriate direction in the cooling process and removed just prior to the measurements of polarization current. Without this poling procedure, an equal population of $\pm P$ domains appears and no polarization current can be observed.

Heat Capacity Measurement

Heat capacity was measured using a thermal relaxation method [4]. We assume that the specimen (with heat capacity C and temperature T) and the thermal bath (with temperature T_0) is connected by the wire (with thermal conductivity k), and thermal power P is provided to the specimen via the heater attached to the sample platform[3] (Fig. 2.1). The sample temperature T as a function of time t obeys the following equation.

$$P(t) = k(T - T_0) + C\frac{\mathrm{d}T}{\mathrm{d}t} \tag{2.7}$$

[3] Here, we assume the good thermal contact between the specimen and the sample stage. For this purpose, Apiezon-N grease is used to fix the sample.

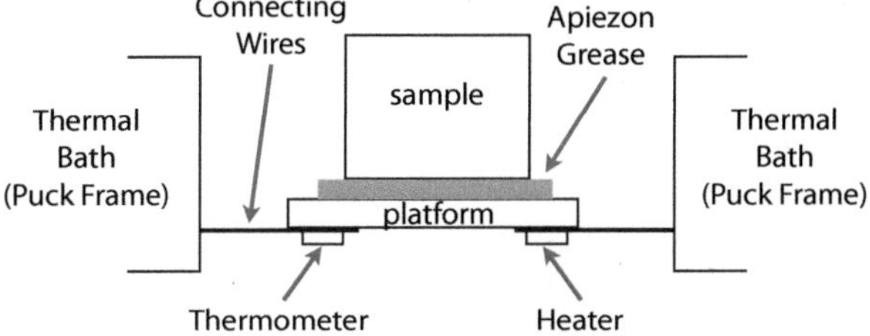

Fig. 2.1 Thermal connections to sample and sample platform [4]

When the heater is switched off (i.e. $P = 0$), the temperature variation between the specimen and heat bath $(T - T_0)$ shows the relaxation behavior.

$$(T - T_0) = (T_1 - T_0)\exp(-t/\tau) \tag{2.8}$$

Here, τ and T_1 represent thermal relaxation time and sample temperature just before the heater is turned off, respectively. The heat capacity C of the specimen can be obtained by the following equation.

$$\tau = C/k \tag{2.9}$$

In reality, C also contains the contribution from the sample platform and thermal grease. Thus, we performed the same measurements with and without the sample, and deduced the pure contribution from the specimen.

Electron Spin Resonance

ESR signal was measured by JEOL JES-FA200 at X-band frequency ($\nu \sim 9.0$ GHz). The experimental setup is illustrated in Fig. 2.2a. The sample sits in a resonant cavity, and microwave radiation with fixed frequency enters via a waveguide. To induce the Zeeman splitting, static magnetic field H is applied perpendicular to the magnetic component of microwave. Electron paramagnetic resonance is observed when the energy gap (Δ) between $m_s = \pm 1/2$ states becomes equal to $h\nu$ (Fig. 2.2b).[4]

$$h\nu = \Delta = g\mu_{\mathbf{B}}H. \tag{2.10}$$

We measured the microwave absorption as a function of H, and deduced the g-value from the magnitude of resonance field. By rotating the specimen within a cavity, the angle-dependence (i.e. anisotropy) of g-value was also investigated.

[4] Here, we assume $CuCl_2$ with $S = 1/2$ as the target compound. To avoid the humidity, the specimen was sealed into a quartz tube filled with Ar-gas.

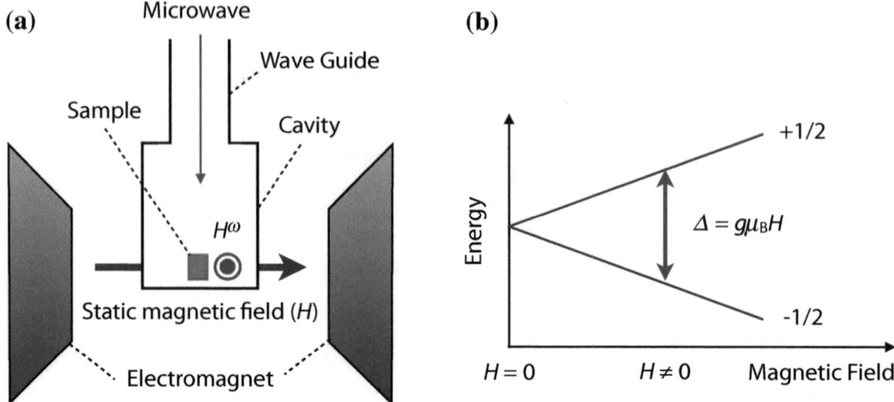

Fig. 2.2 **a** Schematic illustration of the experimental setup for ESR measurement. **b** Zeeman splitting between $m_s = \pm 1/2$ states

Polarized Neutron Scattering

Polarized neutron diffraction study was performed with Polarized Neutron Triple-Axis Spectrometer (PONTA) of ISSP-NSL at JRR-3M. Neutron diffraction is known as one of most powerful experimental methods to determine the magnetic structure in solids. The employment of polarized neutron provides additional sensitivity to the spin-orientation and spin-helicity (clockwise or counter-clockwise manner of spin rotation) of the target compound, which enables more precise characterization of non-collinear magnetic structure.[5]

Figure 2.3 indicates the schematic illustration of experimental setup. We define the scattering vector $\mathbf{Q}$ as $\mathbf{Q} = \mathbf{k}_f - \mathbf{k}_i$, where $\mathbf{k}_i$ and $\mathbf{k}_f$ represent wave vectors of incident and scattered neutrons, respectively. Originally, each incident neutron possesses the random spin direction. Neutron spin orientation ($\mathbf{S}_n$) is aligned by the Heusler (111) monochromator, and the polarized state is sustained by the guide magnetic field applied throughout the neutron beam path. $\mathbf{S}_n$ can be reversed by the "spin-flipper", which is inserted between the specimen and monochromater. The specimen is surrounded by the Helmholtz coil. Since quantization axis of neutron spin is always parallel to the field direction, we can reorient $\mathbf{S}_n$ by controlling the direction of magnetic field ($\sim$10 mT) generated via the Helmholtz coil. Depending on the purpose, we employed two different configurations; $\mathbf{S}_n \perp \mathbf{Q}$ and $\mathbf{S}_n \perp \mathbf{Q}$.

The flipping ratio of polarized to unpolarized neutrons measured at the (2,1,0) nuclear reflection was sufficiently large; 33 for $\mathbf{S}_n \perp \mathbf{Q}$ and 27 for $\mathbf{S}_n \parallel \mathbf{Q}$. The sample was mounted on a sapphire plate in a closed-cycle helium refrigerator, so that the horizontal scattering plane of the spectrometer coincided with the (h k 0) zone. The neutron energy was fixed at 13.47 meV, and only the elastic scattering experiments

[5] For detail, see "LiCu2O2: Correlation Between Spin-Helicity and Electric Polarization Vector".

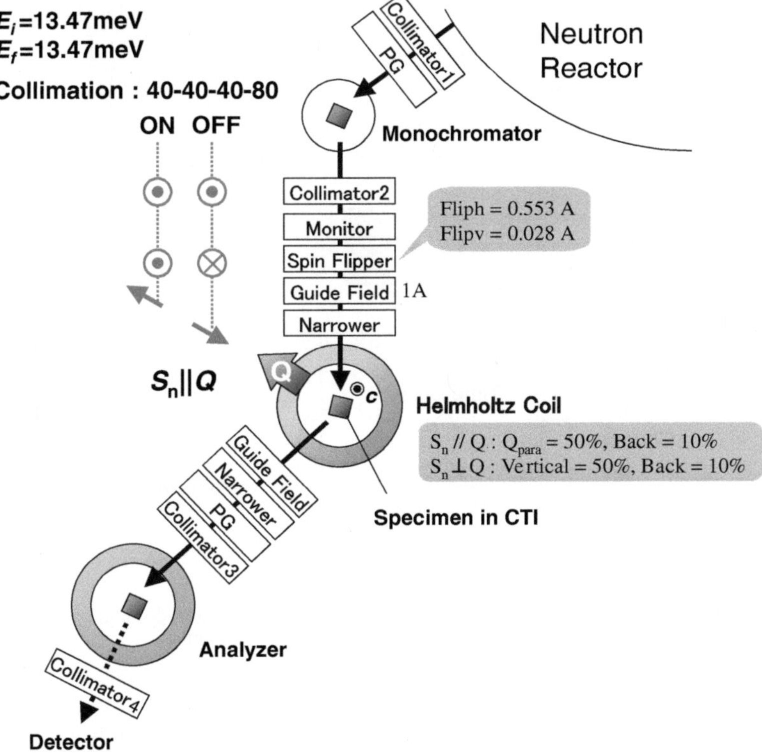

Fig. 2.3 Experimental setup for polarized neutron diffraction study with $\mathbf{S}_n \perp \mathbf{Q}$ condition. *Red arrows* and ON/OFF indicate the spin direction of incident neutron and the state of spin-flipper, respectively

were performed. Pyrolytic graphite (PG) filters were used to eliminate higher harmonic reflections from the monochromator. Collimators ($40'$-$40'$-$40'$-$80'$) were used to obtain the parallel neutron beam. The size of the specimen ($LiCu_2O_2$) used for the neutron study is $12\,mm^2$ (*ab* plane) $\times 0.6\,mm$ (*c*-axis). All the data for $LiCu_2O_2$ presented in this thesis were measured on the identical sample.

THz Time-Domain Spectroscopy

We used the terahertz time-domain spectroscopy (THz-TDS) in a transmission geometry to obtain complex refractive index n without using Kramers-Kronig analysis. THz-TDS can easily access the low-energy electrodynamics in solids, with typical frequency range from 0.1 to 3 THz. The lower limit of the measurable frequencies depends on the size of the samples since 0.1 THz corresponds to 3 mm in wave-

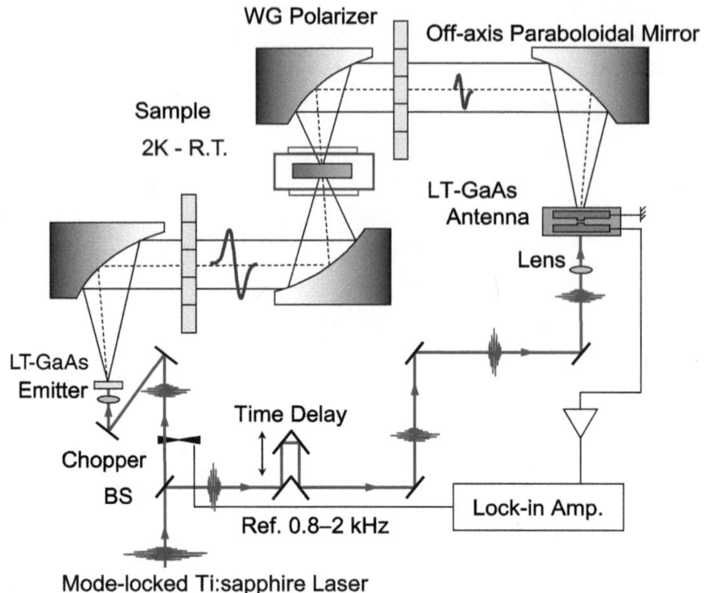

Fig. 2.4 Schematic illustration of the experimental setup for THz time-domain spectroscopy in a transmission geometry. (Adapted with permission from [5], ©2008 APS)

length. The sample was attached to the perforated Cu holder with a diameter of $2.4 \sim 3.5$ mm, which was selected to suitably match the size of the samples.

Figure 2.4 shows the schematic illustration of the experimental setup. For the measurements, we used the photoconducting (PC) sampling technique to obtain terahertz radiation pulse. The femtosecond laser pulses delivered from the mode-locked Ti:sapphire laser were used as a source. The laser pulses were divided by a beam splitter (BS). One was used as the pump and another was used as the gate. The pump pulses were irradiated on the low-temperature-grown GaAs (LT- GaAs) photoconducting device coupled with a dipole antenna to induce the terahertz radiation. The wire grid (WG) polarizers were inserted in between off-axis paraboloidal mirrors to obtain the linear polarization. The polarized terahertz pulse was focused on the sample by off-axis paraboloidal mirrors. The cryostat was placed within the box filled with dry N_2-gas to eliminate the absorption of water. The gate pulses were introduced to the LT-GaAs antenna after the appropriate time delay. The pump pulses were mechanically chopped and the photocurrent induced by the electric field of the terahertz pulse was lock-in detected. Therefore, the induced photocurrent, which depends on both amplitude and phase of terahertz radiation, can be obtained by varying the optical delay line.

In THz-TDS, wave form of irradiated pulse electric field (ranging within a few picoseconds) is directly measured in time domain with and without the specimen. Typical example of measured terahertz wave form was indicated in Fig. 2.5. They are converted into the frequency domain via the fast Fourier transformation (FFT), and

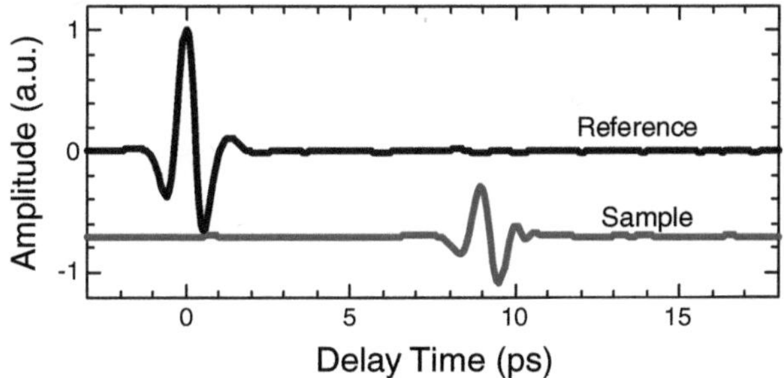

Fig. 2.5 Typical example of the measured terahertz wave form in time-domain with and without the specimen. The amplitude of the transmitted terahertz wave form with the sample was vertically offset for clarity. (Adapted with permission from [5], ©2008 APS)

the spectra of complex transmission constant (t) is deduced using the relationship $t = E^{\omega}_{\text{sample}}/E^{\omega}_{\text{ref}}$. Here, we can avoid the effect of multiple reflections and resultant interference by restricting the time range of FFT. Obtained t is further converted into complex refractive index $n = \sqrt{\epsilon\mu}$ using the following relationship;

$$t = \frac{2\mu}{n+\mu}\frac{2n}{n+\mu}\exp\Big[-i\frac{\omega}{c}d(n-1)\Big], \quad (2.11)$$

where ϵ, μ, d, ω, and c represent the complex dielectric constant, complex magnetic permeability, sample thickness, frequency of light, and velocity of light, respectively. To numerically solve Eq. (2.11), we approximate the pre-exponential factor by $4n/(n+1)^2$ assuming $\mu \simeq 1$ unless otherwise noted. As shown later, this approximation hardly affects the obtained n (or $\epsilon\mu$) spectrum (see "Electromagnon in the Paraelectric Collinear Spin State ").

References

1. Zhao TR, Hasegawa M, Takei H (1996) Crystal growth and characterization of cuprous ferrite ($CuFeO_2$). J Cryst Growth 166:408
2. Park S, Choi YJ, Zhang CL, Cheong S-W (2007) Ferroelectricity in an $S = 1/2$ chain cuprate. Phys Rev Lett 98:057601
3. Maljuk A, Kulakov AB, Sofin M, Capogna L, Strempfer J, Lin CT, Jansen M, Keimer B (2004) Flux-growth and characterization of $NaCu_2O_2$ single crystals. J Cryst Growth 263:338
4. Physical Property Measurement System Heat Capacity Option User's Manual, Quantum Design Inc
5. Kida N, Ikebe Y, Takahashi Y, He JP, Kaneko Y, Yamasaki Y, Shimano R, Arima T, Nagaosa N, Tokura Y (2008) Electrically driven spin excitation in the ferroelectric magnet $DyMnO_3$. Phys Rev B 78:104414

Chapter 3
Magnetoelectric Response in Triangular Lattice Antiferromagnets

Introduction

As mentioned in "Magnetic Frustration", triangular lattice antiferromagnet is one of the simplest and most typical examples of frustrated spin system [1]. When we assume classical Heisenberg spins and consider only the nearest neighbor interaction, 120°-spin order becomes magnetic ground state. However, depending on the strength of next-nearest neighbor (or more distant) interaction and magnetic anisotropy, more complex magnetic order can also be realized. The purpose of this chapter is to investigate the magnetoelectric response of various types of magnetic order in triangular lattice.

AMO_2 and MX_2

In this chapter, we mainly investigate triangular lattice antiferromagnets with two-types of chemical composition; AMO_2 and MX_2 (Figs. 3.1 and 3.2). In both cases, each element forms two-dimensional triangular lattices independently. These layers stack along the c-axis in order of -(O-M-O)-A-(O-M-O)-A- for AMO_2, and -(X-M-X)-(X-M-X)- for MX_2. Here, the stacking manner within (O-M-O) or (X-M-X) is always rhombohedral. Thus, all compounds possess a common structural unit of MO_2 or MX_2 (Fig. 3.3b), which consists of a layer of transition metal ion (M) sandwiched by layers of ligand anion (oxygen or halogen (X = Cl, Br, I)). While the choice of A, M and X ions causes several different manners (but with the same order) of stacking, the site symmetry of magnetic M-ion is always $\bar{3}m$ i.e. centrosymmetric. Since intra-plane magnetic interaction is much larger than inter-plane interaction, they can be considered as the quasi-two dimensional spin system.

Note that each magnetic M-ion is surrounded by the octahedra of ligand ions (Fig. 3.3a). The crystal field is expected to split $3d$-orbital of M-ion into high-energy twofold e_g states and low-energy threefold t_{2g} states. Thus, in case of d^9 or d^4 electron

S. Seki, *Magnetoelectric Response in Low-Dimensional Frustrated Spin Systems*, Springer Theses, DOI: 10.1007/978-4-431-54091-5_3,

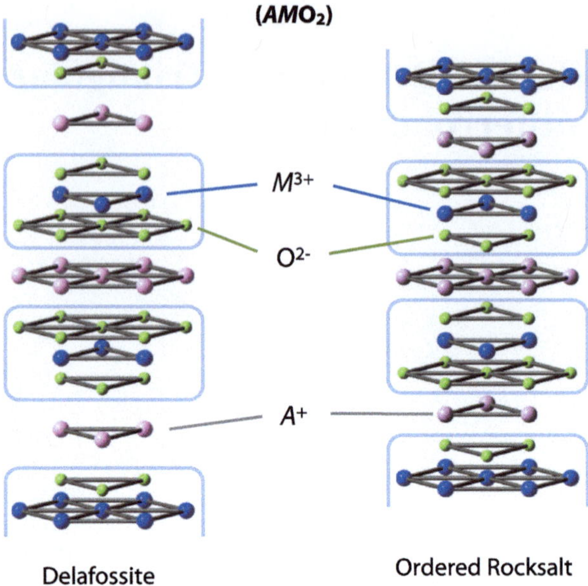

Fig. 3.1 Crystal structure of AMO_2 compounds

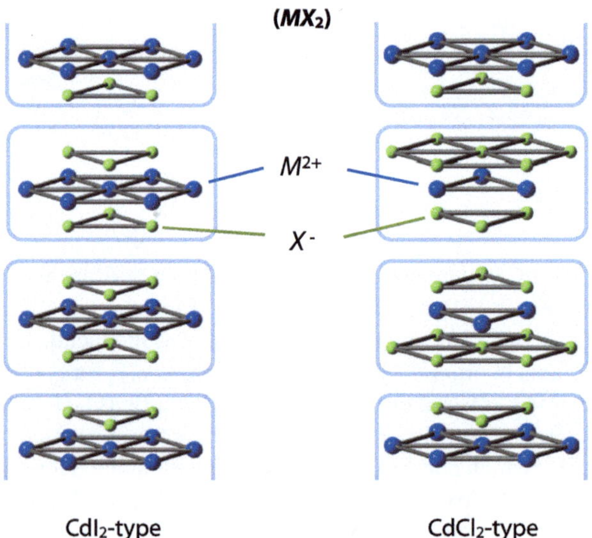

Fig. 3.2 Crystal structure of MX_2 compounds

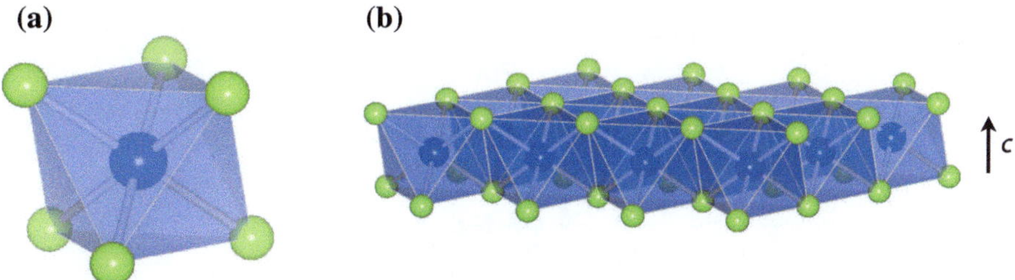

Fig. 3.3 **a** Magnetic ion surrounded by the octahedra of ligand anion. **b** Schematic view of (O-M-O) or (X-M-X) structural units

configuration, Jahn-Teller distortion can take place to lift the degeneracy of e_g states and reduce the total energy.[1] In this chapter, we only treat the compounds without such Jahn-Teller distortion.

Ferroelectricity Induced by Proper Screw Magnetic Order in CuFeO₂

$CuFeO_2$ is known as the first example of triangular lattice helimagnets with magnetically-induced ferroelectricity [2]. This compound is characterized by delafossite crystal structure with space group $R\bar{3}m$ [3], and magnetic moment is carried by Fe^{3+} ion with $S = 5/2$. $CuFeO_2$ undergoes subsequent magnetic phase transitions under magnetic field applied along the c-axis (Fig. 3.4), keeping magnetic propagation vector $\vec{k} = (q, q, 3/2)$ [4]. For simplicity, we define the q-vector ($\vec{q}$) as the in-plane component of magnetic propagation vector. At zero magnetic field, a 4-sublattice collinear $\uparrow\uparrow\downarrow\downarrow$ magnetic ground state (CM4) with commensurate $q = 0.25$ and spin direction along the c-axis is realized below 11 K [5]. With 7 T $< H <$ 12 T, the CM4 phase is replaced by noncollinear proper screw magnetic phase (NC) with incommensurate $q \sim 0.202$, where spin rotates within a plane perpendicular to the q-vector (Fig. 3.5c) [6]. Above 12 T, a 5-sublattice collinear $\uparrow\uparrow\uparrow\downarrow\downarrow$ structure (CM5) with commensurate $q = 0.20$ and spin direction along the c-axis appears [4]. The $H-T$ phase diagram for $CuFeO_2$ is summarized in Fig. 3.4. Starting from either of these magnetic ground states, increase of temperature (T) first induces sinusoidally-modulated collinear (ICM1) magnetic phases with incommensurate $q \sim 0.22$ [7] and then produces a paramagnetic (PM) phase.

In 2006, Kimura et al. first reported the emergence of ferroelectricity in the proper screw magnetic phase of $CuFeO_2$ [2]. Here, the P in the NC phase appears parallel to q-vector (Fig. 3.5c) [8, 9]. This multiferroic behavior of $CuFeO_2$ is rather

[1] This distortion makes $CuCl_2$ (with d^9) a quasi-one dimensional spin system. The magnetoelectric response of $CuCl_2$ is discussed in "$CuCl_2$".

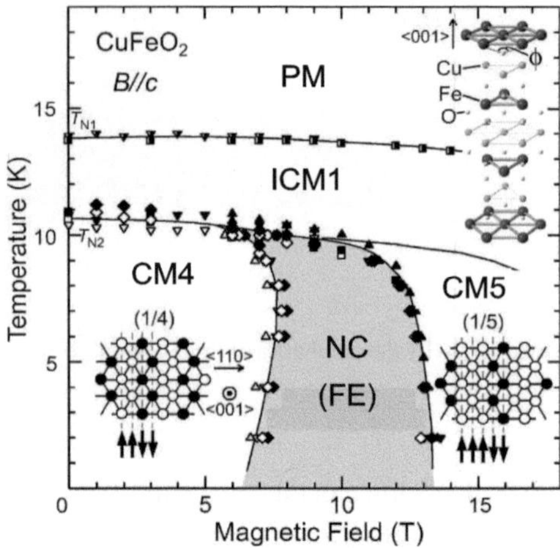

Fig. 3.4 $H-T$ phase diagram of $CuFeO_2$ with H applied along the c-axis. *Open* and *filled symbols* represent the data points in the cooling (or H-decreasing) and warming (or H-increasing) runs, respectively. *Diamond*, *square*, *triangle*, and *inverse triangle* data points were obtained by measurements of magnetization, dielectric constant, electric polarization, and magnetostriction, respectively. Inset shows schematic illustration of magnetic structures on Fe^{3+} sites at (*left*) CM4 and (*right*) CM5 states. *White* and *black circles* correspond to *up* and *down spin states*, respectively. Ferroelectricity (FE) is observed in the noncollinear proper screw (NC) magnetic phase (*gray area*). (Adapted with permission from [2], ©2006 APS.)

unexpected, since the inverse D–M model predicts that proper screw magnetic order can induce no ferroelectric polarization. The spacial average of electric dipole moment induced by exchange striction mechanism is also zero for the incommensurate magnetic order on centrosymmetric lattice. Nevertheless, the present emergence of ferroelectricity can be justified from the viewpoint of symmetry. As already mentioned in "Magnetically-Induced Ferroelectricity", a proper screw magnetic structure retains the following symmetry elements: $2'$-axis along the q-vector as well as 2- and $2'$-axes perpendicular to the q-vector (Fig. 3.5a). Since orthogonal arrangement of the $2'$- and 2-axes prevents the system from being polar, a proper screw magnetic order alone cannot induce ferroelectricity in general. If the crystal lattice has no 2-axis symmetry within a plane perpendicular to the q-vector, however, only $2'$-axis along the q-vector can remain unbroken and the appearance of ferroelectricity with $P \parallel q$ is allowed (Fig. 3.5c). All crystallographic symmetry elements at magnetic Fe^{3+} site are indicated in Fig. 3.5b, and we can see the crystal structure of $CuFeO_2$ does satisfy the above condition for the emergence of electric polarization [9].

While the microscopic origin of such magnetoelectric coupling is yet to be clarified, recently Jia et al. suggested that the spin-dependent modulation of covalency (hybridization) between metal d-state and ligand p-state [10], which also stems from the spin-orbit interaction, can induce the finite local polarization along the metal-ligand bond direction. While this term often oscillates and cancels out within the crystal, Arima recently suggested the finite $\vec{P} \parallel \vec{q}$ component can survive on the

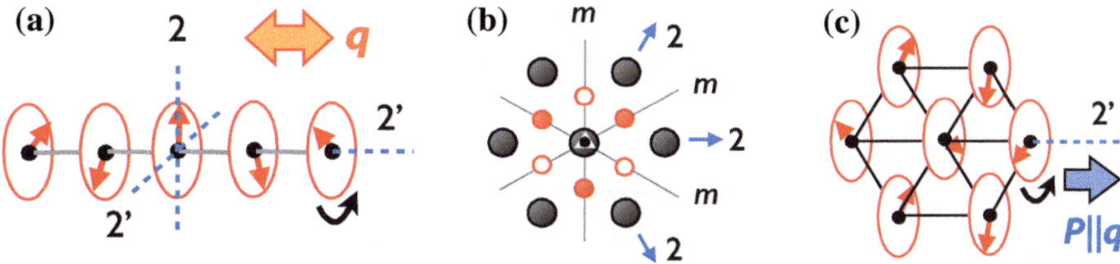

Fig. 3.5 **a** Symmetry elements compatible with proper screw magnetic structure. **b** Symmetry elements at magnetic Fe^{3+} site (*large black circle*) with crystallographic site symmetry $\bar{3}m$: two fold rotation axis (2), reflection mirror (m), and threefold rotation axis along the c-axis with inversion center (*triangle* with *small circle*). O^{2-} site *above* (*below*) the Fe^{3+} layer is indicated as *small red closed* (*open*) circle. (c) Proper screw magnetic order on delafossite crystal lattice. Directions of induced electric polarization (P) and magnetic q-vector are also indicated

delafossite lattice with proper screw magnetic order [9]. This model naturally predicts that the reversal of P-vector is coupled with the reversal of vector spin chirality (i.e. clockwise or counter-clockwise manner of spin rotation), which was later confirmed by the polarized neutron scattering experiments [8].

Overview

As suggested in the case of $CuFeO_2$, the symmetry of triangular lattice often allows the appearance of magnetically-induced ferroelectricity which cannot be explained by either inverse D–M mechanism or exchange striction. In the following, we investigate the magnetoelectric response of $CuFeO_2$ in detail from the viewpoint of the impurity-doping effect ("Impurity-Doping-Induced Ferroelectricity"), domain control ("Magnetic Digital Flop of Ferroelectric Domain"), and dynamics ("Electromagnon in the Paraelectric Collinear Spin State"). In "$ACrO_2$: Ferroelectricity Induced by 120°-Spin Order" ", we study the magnetoelectric response of $ACrO_2$ (A = Cu, Ag, Li, and Na) with 120°-spin order, which is also predicted to induce no ferroelectric polarization according to the inverse D–M scheme. The results obtained in AMO_2 compounds are further utilized for the investigation and interpretation of ME behavior in MX_2-type halides ("MX_2-Type Halides with CdI_2 Structure"").

$CuFeO_2$

Impurity-Doping-Induced Ferroelectricity

Introduction

In the frustrated spin system, the competition of interactions often leads to variety of similar low energy states which almost degenerate with the magnetic ground states.

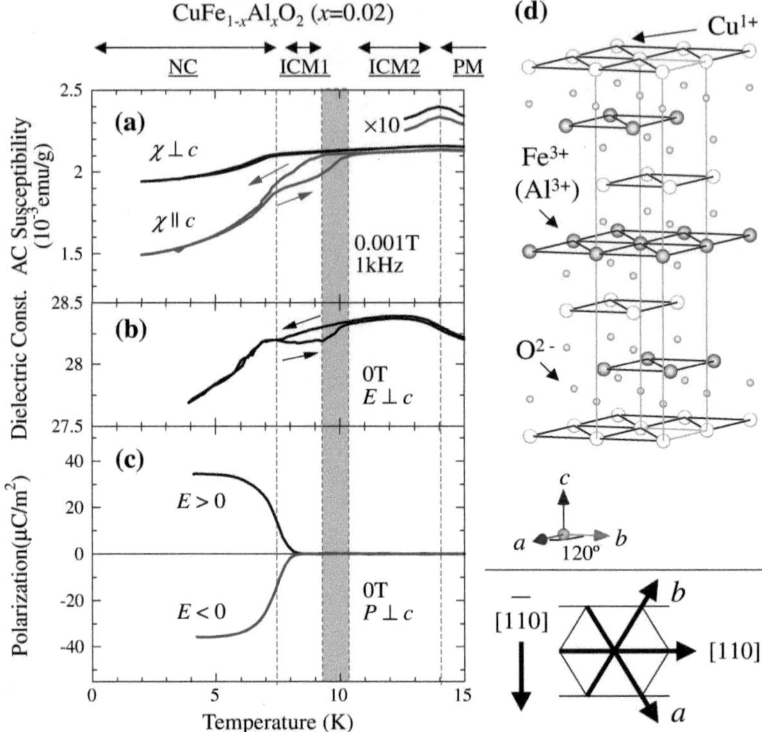

Fig. 3.6 Temperature profiles of **a** ac susceptibility measured with magnetic fields parallel and perpendicular to the c-axis (magnified data is also plotted separately), **b** in-plane dielectric constant, and **c** electric polarization perpendicular to the c-axis for $CuFe_{1-x}Al_xO_2$ ($x = 0.02$). The *arrows* in **a** and **b** indicate the thermal scan direction of the measurement. Two opposite poling electric fields are used for the polarization measurement ($E > 0$ and $E < 0$). **d** Schematic crystal structure of delafossite $CuFe(Al)O_2$

In such a situation, even a small perturbation alters the relative stability of these competing magnetic states; The H-induced successive magnetic phase transitions in $CuFeO_2$ is a typical example. Likewise, the substitution of magnetic Fe^{3+} with non-magnetic ion is also expected to significantly affect its magnetic phase diagram. In this section, we investigate such a dilution effect on the magnetism and magnetoelectric response for $CuFeO_2$.

Results

In the following, we focus on the case of Al^{3+}-doping on Fe^{3+} site. The magnetism of $CuFe_{1-x}Al_xO_2$ ($x = 0.02$) has previously been investigated [11], although no dielectric measurements has been performed for Al-doped specimens. Figure 3.6 shows the temperature dependence of the AC susceptibility, dielectric constant, and polarization for $CuFe_{1-x}Al_xO_2$ (x = 0.02). There are known three antiferromagnetically ordered phases at zero magnetic field in this crystal. [11] While the wave vector of the magnetic modulation is along [110] in all the phases, the direction of the magnetic moment is different among the phases. As the temperature decreases, a sinusoidal and collinear magnetic structure shows up at 14 K (ICM2). The magnetic moments are canted by 50° from the c-axis in the ICM2 phase [12], whereas they are aligned to the c-axis in the lower temperature phase below 11 K (ICM1) [7]. The noncollinear magnetic structure (NC) is realized below 7 K. A broad peak of susceptibility around 14 K for both $H \parallel c$ and $H \perp c$ corresponds to the transition from paramagnetic phase (PM) to ICM2. A kink is observed at 9 K in a cooling run, and at 10 K in a warming run in the c-axis susceptibility. These correspond to the transition from ICM2 to ICM1 [13]. The variation of the transition temperature due to the hysteresis is indicated by the width of the gray bar in Fig. 3.6. Kinks to separate ICM1 and NC are also found at 7 K for both $H \parallel c$ and $H \perp c$. The difference in the anisotropy of susceptibility between ICM1 and ICM2 is due to the direction of the ordered spins. The dielectric constant shows kinks at 7 K, similarly to the magnetic susceptibility. The hysteresis is also observed for the dielectric constant between 7 and 10 K. These suggest the strong coupling between the electric and magnetic properties in this material. Most importantly, the spontaneous polarization begins to increase at 7 K, which is the transition temperature of the noncollinear phase. We confirm the ferroelectric nature, that is the sign reversal of the polarization when the opposite poling field is used. This polarization is observed even at zero magnetic field, in contrast with the case for x = 0.00 [2], where polarization is induced only when magnetic fields of 6–13 T are applied along the c-axis.

Hereafter, we investigate the systematic evolution of the magnetoelectric phase with the variation of Al-doping x. The magnetization curves for x = 0.00 at various temperatures are shown in Fig. 3.7a. Below 9 K, two notable steps are discerned in each curve. The step at the lower field corresponds to the transition from the commensurate 4 sublattice (CM4: ↑↑↓↓ with collinear spin directions along the c-axis) to the noncollinear magnetic structure. The higher-lying one is due to the emergence of the commensurate 5 sublattice (CM5: ↑↑↑↓↓ with collinear spin directions along the c-axis) magnetic structure. The wave vectors of both CM4 and CM5 are also parallel to (110). Figure 3.8a depicts the temperature dependence of electric polarization at various magnetic fields for x = 0.00. The spontaneous polarization is observed between 7 and 12 T. This magnetic field region corresponds to the noncollinear magnetic phase. The results for x = 0.00 are consistent with the previous report [2].

In the case for x = 0.01 (Fig. 3.7b), the two magnetic phase transitions are also observed below 7 K, while the transition field from CM4 to NC considerably decreases. The larger magnetic-field hysteresis of the transition is observed than for x = 0.00. Above 6.8 K, the magnetization is linear with the magnetic field below 7 T in a field-decreasing run, while the step-like structure is observed around 4 T in a field-increasing run (for the detail of the cooling procedure, see the figure caption). This suggests that NC phase exists even at zero field in this temperature region once after high enough field is applied. In Fig. 3.8b, we show the temperature dependence

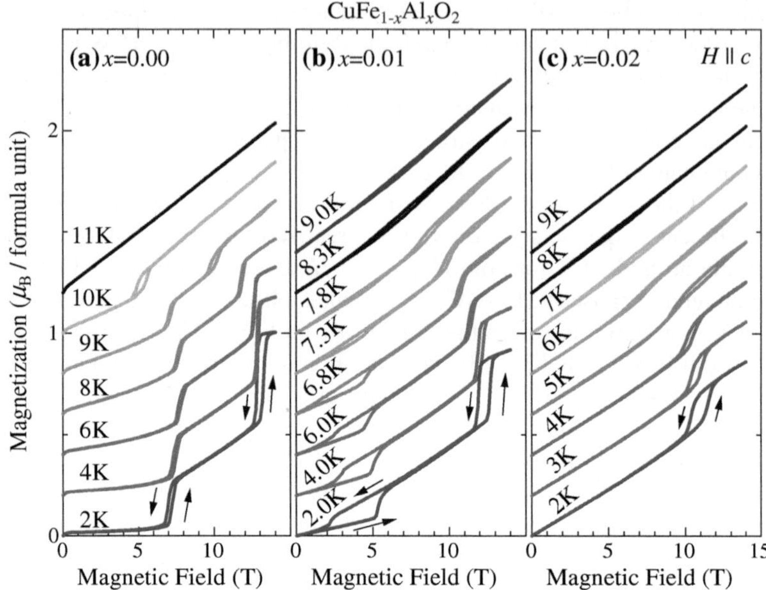

Fig. 3.7 Magnetic field dependence of magnetization at various temperatures for $CuFe_{1-x}Al_xO_2$ with **a** $x = 0.00$, **b** $x = 0.01$, and **c** $x = 0.02$. *Arrows* indicate the field-scan direction of the measurement. In the measurement for $x = 0.01$, the sample was warmed up to 20 K and cooled down without magnetic field, prior to each field-increasing run

of the polarization at various fields for $x = 0.01$. At zero field, the finite spontaneous polarization is observed between 6 and 9 K when the temperature is cooled without entering CM4 phase (Here, the poling electric field (∼200 kV/m) was removed after entering the ferroelectric phase, and then the displacement current was measured with both increasing and decreasing temperature from the stopped temperature). On the other hand, once after cooling down below 6 K, only negligible polarization can be observed in a warming process. A similar behavior is observed below 3 T.[2] Above 4 T, there is a spontaneous polarization even at the lowest temperature. The polarization disappears at the NC-to-CM5 transition field.

Figure 3.7c depicts magnetization curves for $x = 0.02$. Only a single step is observed around 10 T below 4 K, as caused by the transition from the NC to CM5 phase. This confirms that the CM4 phase is completely suppressed for the $x = 0.02$ doping. Figure 3.8c shows the temperature dependence of polarization for $x = 0.02$. The polarization begins to increase around 7 K with decreasing temperature at zero magnetic field. The reentrant paraelectric behavior as observed for $x = 0.01$ at $H < 3$ T is not observed for the case of $x = 0.02$. The onset temperature of the spontaneous polarization decreases with magnetic field. The ferroelectric behavior disappears

[2] To be accurate, a small polarization remains even below the transition temperature of CM4 phase for $x = 0.01$ and increases with magnetic field. This is likely due to the phase coexistence of the CM4 and NC arising from the first-order transition nature [14].

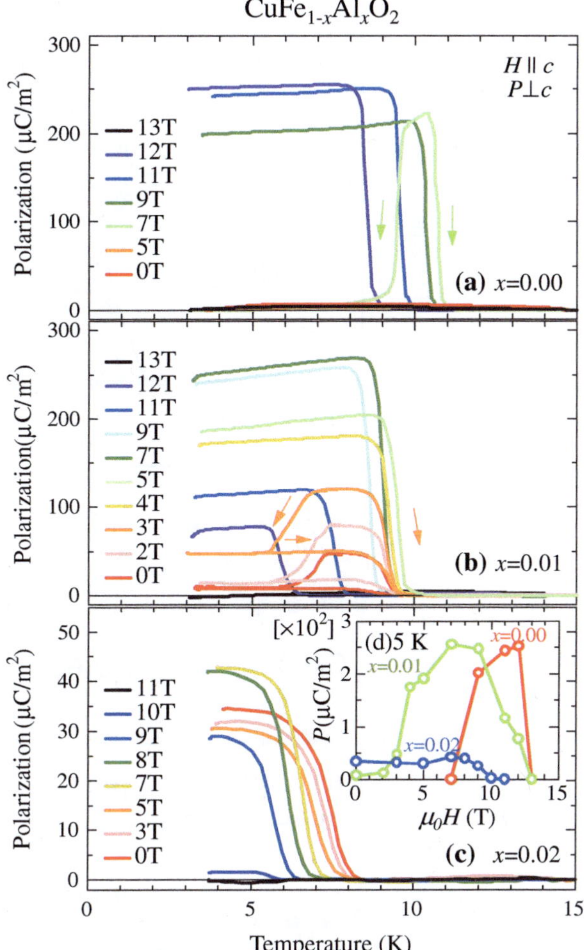

Fig. 3.8 Temperature dependence of electric polarization perpendicular to the *c*-axis at various magnetic fields for $CuFe_{1-x}Al_xO_2$, **a** $x = 0.00$, **b** $x = 0.01$, **c** $x = 0.02$. *Arrows* in **a** and **b** indicate the thermal scan direction of the measurement. After a proper poling procedure with $E \sim 200$ kV/m, each measurement was made in a warming run unless indicated by the *arrows* (see text). **d** Magnetic field dependence of polarization at 5 K obtained from the above scans

above 12 T. Thus, the ferroelectric region seems to coincide with the region of NC phase also for the Al-doped crystals. In the inset of Fig. 3.8d, we plot the magnetic field dependence of polarization at 5 K for all the crystals. The onset field of ferroelectric or NC phase decreases dramatically with the impurity doping.

The variation of these magnetoelectric states with magnetic field and Al-doping are clearly figured out by the phase diagram as shown in Fig. 3.9; the temperature versus magnetic field (parallel to the *c*-axis) phase diagrams for (a) $x = 0.00$, (b) $x = 0.01$, and (c) $x = 0.02$. They were determined by the measurements of magnetization and dielectric constant. The phase diagrams for $x = 0.00$ and $x = 0.02$, apart from the identification of the ferroelectric state, agree with the previously reported ones determined by neutron scattering [4, 11]. Although the neutron scattering was done only under zero magnetic field for $x = 0.01$ [15], each magnetic phase can be

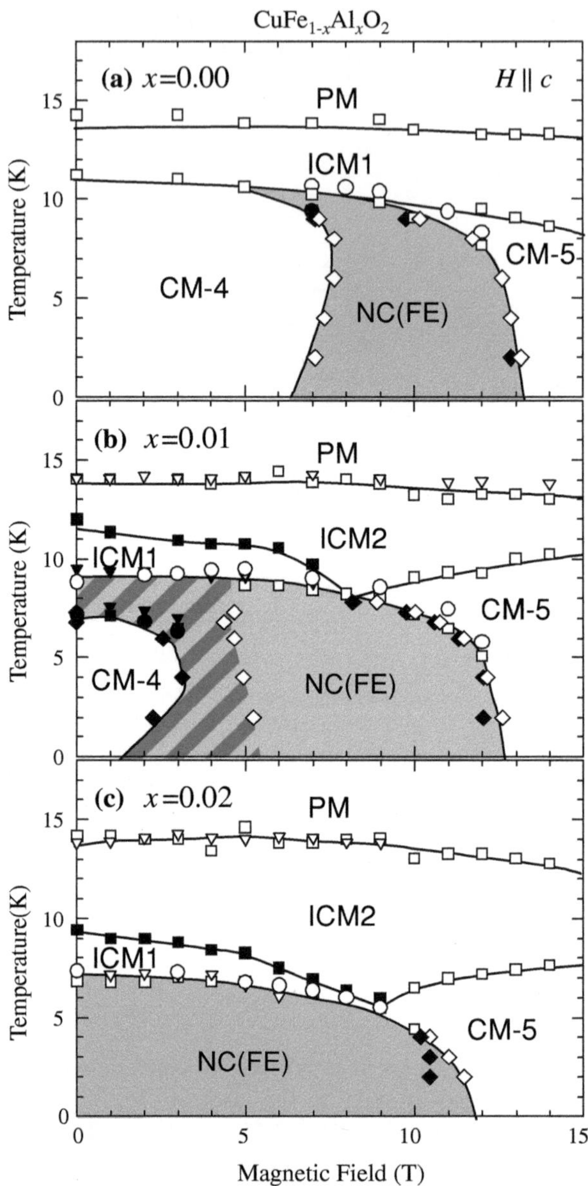

Fig. 3.9 Temperature (T) versus magnetic field (H) phase diagram of $CuFe_{1-x}Al_xO_2$ for **a** $x = 0.00$, **b** $x = 0.01$, and **c** $x = 0.02$ ($H \parallel c$). *Circle*, *triangle*, *square*, and *diamond* data points were obtained by measurements of electric polarization, dielectric constant, magnetization (T-dependence), and magnetization (H-dependence), respectively. *Open* and *filled symbols* represent the anomalies in the increasing and decreasing T or H runs, respectively. Spontaneous electric polarization, i.e. the ferroelectricity (FE) was observed in the noncollinear spin order (NC) state (the *shadowed area*). In the *hatched area*, the magnetic structure depends on the hysteresis. The CM4 phase remains in a field-increasing or temperature-increasing run, otherwise the NC phase shows up in this area

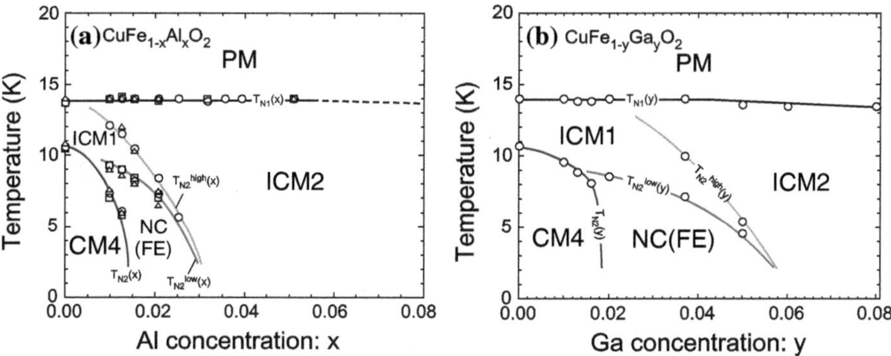

Fig. 3.10 $x-T$ phase diagrams for **a** $CuFe_{1-x}Al_xO_2$ and **b** $CuFe_{1-x}Ga_xO_2$ under zero magnetic field. (Adapted with permission from [16], ©2009 IOP.)

specified by the comparison with the phase diagrams for $x = 0.00$ and $x = 0.02$. The ICM2 phase appears when Al is doped into $CuFeO_2$. The critical field for the transition from CM4 to ferroelectric NC decreases dramatically but systematically with Al doping. The CM4 phase is completely suppressed for $x = 0.02$. The onset of the spontaneous polarization is also plotted with circles. Importantly, in each phase diagram, spontaneous polarization is observed only in the NC phase (shadowed region), not in the collinear nor paramagnetic phase. This indicates the inseparable relation between the noncollinear spin structure and ferroelectricity.

Discussion

As demonstrated above, even a small amount (~2 %) of Al-doping can drastically change the magnetic (and dielectric) phase diagram of $CuFeO_2$. This also reduces the critical magnetic field necessary for the induction of ferroelectric helimagnetic phase down to 0 T.

After the report of the present study, similar dilution effect has been investigated using non-magnetic Ga^{3+}-ion [18]. In Fig. 3.10, the $x - T$ phase diagrams at zero magnetic field for $CuFe_{1-x}Al_xO_2$ and $CuFe_{1-x}Ga_xO_2$ are indicated [16]. While the change of magnetic phase relationship is more moderate in case of Ga-doping, ferroelectric proper screw magnetic phase (NC(FE)) is stabilized down to 0 T even for $CuFe_{1-x}Ga_xO_2$ ($x > 0.02$). The difference between Al-doping and Ga-doping is more pronounced in their poling electric field (E_p) dependence of P-value (Fig. 3.11) [17]. Al-doping in $CuFeO_2$ largely enhances the magnitude of E_p necessary for the saturation of P-value (E_p^s), while Ga-doping gives almost no impact on the E_p^s. Since ionic radius of Fe^{3+} (0.65 Å) is much closer to that of Ga^{3+}(0.62 Å) rather than Al^{3+} (0.53 Å), larger local lattice distortion is expected in case of Al-doping. Such a distortion can act as the pinning center for multiferroic domain walls, and thus reduces the sensitivity of P-domain distribution for applied external electric field. This scenario also well explains the seemingly reduced P-value in Al-doped

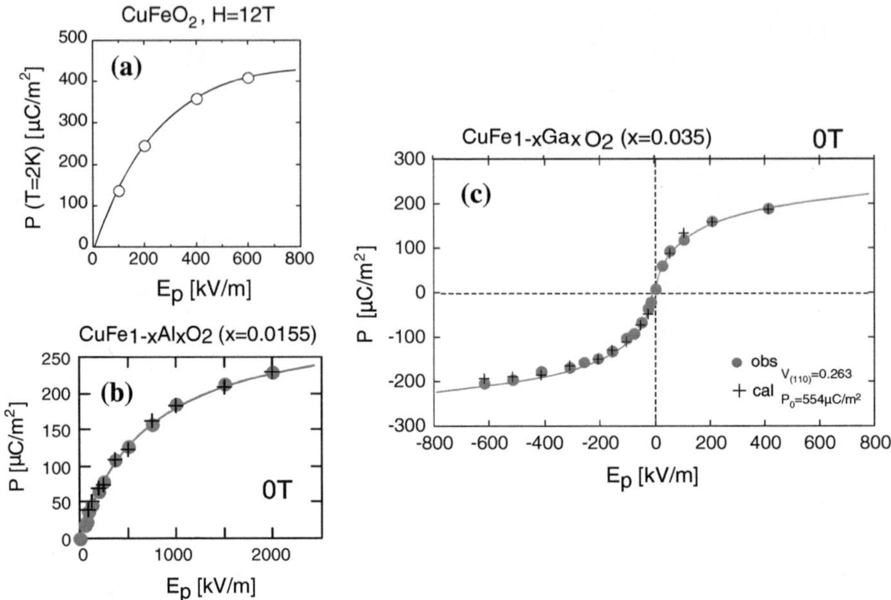

Fig. 3.11 **a–c** Poling electric field (E_p) dependence of electric polarization (P) for $CuFeO_2$, $CuFe_{1-x}Al_xO_2$ ($x = 0.0155$) and $CuFe_{1-x}Ga_xO_2$($x = 0.035$), respectively. (Adapted with permission from [17], ©2009 APS.)

specimen, which is observed under the relatively weak poling electric field $E \sim 200$ kV/m (Fig. 3.8d).

Another important issue is the reason why nonmagnetic impurity can stabilize the ferroelectric proper screw phase. To resolve this problem, first we have to understand the underlying magnetic interactions in $CuFeO_2$. Based on the fitting of spin-wave dispersion obtained by inelastic neutron diffraction study [20], recent theoretical analysis suggested that the magnetism in $CuFeO_2$ is described using the spin Hamiltonian

$$\mathcal{H} = -\frac{1}{2}\sum_{i \neq j} J_{ij}\mathbf{S}_i \cdot \mathbf{S}_j - D\sum_i S_{iz}^2 - g\mu_B H \sum_i S_{iz} \tag{3.1}$$

with $J_1 S = -1.14$ meV, $J_2 S = -0.50$ meV, $J_3 S = -0.65$ meV, $J_z S = -0.33$ meV, and $DS = 0.17$ meV [21]. Here, $J_{ij} = J_1, J_2, J_3$, and J_z depends on the relative vector $\mathbf{r}_i - \mathbf{r}_j$ between sites i and j. Takagi et al. performed the Monte Carlo simulations considering only J_1, J_2, and J_3 in the limit of large D (i.e. Ising spin), and deduced the magnetic ground state as a function of $J_2/|J_1|$ and $J_3/|J_1|$ [22]. According to their calculation, the present exchange parameters for $CuFeO_2$ actually places at the region of 4-sublattice $\uparrow\uparrow\downarrow\downarrow$ spin order [21].

Notably, the spin-wave dispersion of $CuFeO_2$ has the two energy dips at $\mathbf{k}_1$ =(0.21, 0.21, 3/2) and $\mathbf{k}_2$ =(0.29, 0.29, 3/2) in symmetric positions around the commensurate

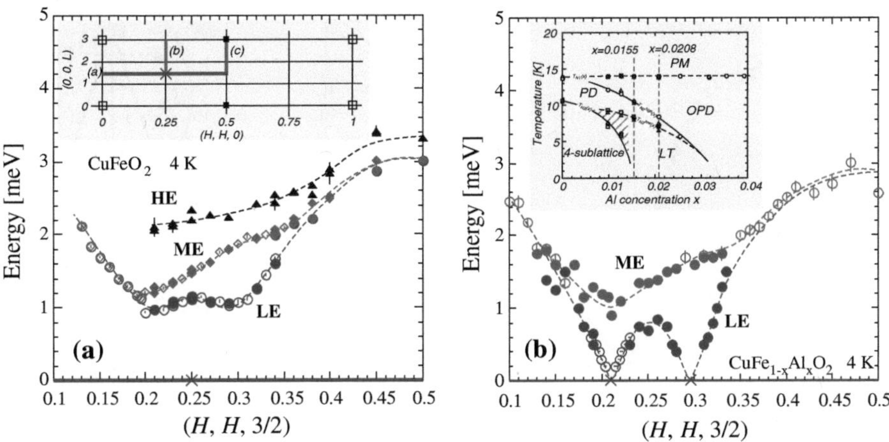

Fig. 3.12 Experimentally determined spin-wave dispersion along the $(h, h, 3/2)$ direction for **a** $CuFeO_2$ with the CM4 magnetic state and **b** $CuFe_{1-x}Al_xO_2$ $(x = 0.02)$ with the ferroelectric NC magnetic state. (Adapted with permission from [19], ©2007 IOP.)

magnetic Bragg point (0.25, 0.25, 3/2) characteristic of the CM4 magnetic ground state (Fig. 3.12a). These $\mathbf{k}_1$ and $\mathbf{k}_2$ are precisely the same wave vectors associated with the H-induced ferroelectric helimagnetic order. Since spin-wave modes in any antiferromagnet linearly split in terms of applied magnetic field, a critical magnitude of magnetic field (H_c) will close the spin gap Δ and destroy the local stability of the CM4 phase; For the observed $\Delta = 0.9$ meV this critical value is $H_c = \Delta / g\mu_B = 7.7$ T, which is in agreement with the experimental observation that helimagnetic magnetic order appears above 7 T [20, 21].

In Fig. 3.12b, the spin-wave dispersion of $CuFe_{1-x}Al_xO_2$ $(x = 0.02)$ determined by the inelastic neutron diffraction study is indicated [19]. Here, the ground state is the ferroelectric helimagnetic state, and we can see that the spin gap at $\mathbf{k}_1$ and $\mathbf{k}_2$ is actually closed in this phase. Fishman et al. suggested that the observed spin-wave dispersion with closed spin gap for $CuFe_{1-x}Al_xO_2$ $(x = 0.02)$ can be reproduced when we reduce the strength of single-ion anisotropy from $DS = 0.17$ meV of $CuFeO_2$ to $DS = 0.12$ meV (Fig. 3.13) [21]. Their Monte-Carlo simulation also indicates that such reduction of D-value stabilizes the proper-screw like magnetic ground state [23]. Therefore, the main effect of nonmagnetic-impurity doping may be to surpass the single-ion anisotropy while keeping the other exchange parameters relatively unchanged. In general, Fe^{2+} (d^4) ion under trigonal crystal field retains the finite orbital angular momentum $|l| = 1$ and host strong Ising anisotropy via the spin–orbit interaction $\lambda(\mathbf{l} \cdot \mathbf{s})$, while Fe^{3+} (d^5) ion can host no magnetic anisotropy [24]. If $CuFeO_2$ contains a considerable number of Fe^{2+} impurities and the Al^{3+} or Ga^{3+} dopants preferentially replace the Fe^{2+} rather than the Fe^{3+} ions, they would have the effect of decreasing the anisotropy D. Interestingly, recent resonant x-ray diffraction experiment on $CuFeO_2$ suggested the existence of charge disproportion

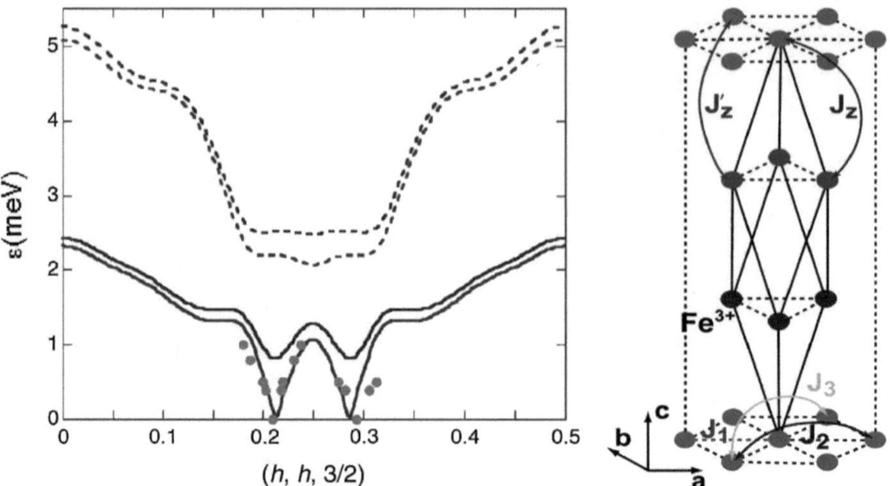

Fig. 3.13 The calculated spin-wave dispersion along the $(h, h, 3/2)$ direction. The *solid* and *dashed curves* denote the two spin-wave branches for the exchange parameters assumed for $CuFeO_2$. The *upper solid* and *dashed curves* are for anisotropy $DS = 0.17$ meV and the *lower* for $DS = 0.12$ meV. The experimentally measured spin-wave frequencies for $CuFe_{1-x}Al_xO_2$ ($x = 0.02$) are plotted as the *solid points*. (Adapted with permission from [21], ©2008 AIP.)

$2Fe^{3+} \rightarrow Fe^{(3+\delta)+} + Fe^{(3-\delta)+}$ with $\delta \sim 0.26$ in the CM4 phase [25]. While the origin of the charge disproportion and its relevance to magnetic anisotropy is not clear at this stage, this discovery may offer an important footstep to fully understand the reason why nonmagnetic impurity doping in $CuFeO_2$ can stabilize the ferroelectric helimagnetic phase.

Conclusion

In this section, we investigated the effect of nonmagnetic impurity doping on the magnetic Fe site for triangular lattice antiferromagnet $CuFeO_2$ with magnetic frustration. Reflecting its competing magnetic ground states, even a small amount of nonmagnetic impurity drastically changes the magnetic phase diagram. Especially, the critical magnetic field necessary for the induction of ferroelectric helimagnetic phase decreases down to zero with Al-doping up to $x = 0.02$. These results prove that not only magnetic field but also site-dilution can often stabilize magnetic ferroelectricity via the modification of spin frustration. Since all magnetically-induced ferroelectrics are reported to contain the magnetic frustration, nonmagnetic impurity doping would be a promising way to tune the magnetoelectric response in such multiferroic materials.

Magnetic Digital Flop of Ferroelectric Domain

Because of the symmetry of triangular lattice, $CuFe_{1-x}Ga_xO_2$ hosts six equivalent multiferroic domains with different P-directions as shown in Fig. 3.14j. In the following, we attempt to control these domains using in-plane H.

Introduction

To obtain the magnetoelectric response via the phase transition, mainly two approaches are possible. One is the H-induced modulation of spin structure in magnetically-induced ferroelectrics, which always leads to the change of direction and/or magnitude of P-vector. Typical examples are the H-induced 90°-flop of spin–spiral plane and associated P-direction observed in $TbMnO_3$, and the induction of ferroelectric helimagnetic phase in $CuFeO_2$ under applied H. However, such directional change or induction of P is not persistent after the removal of H, since original spin structure is generally recovered under zero magnetic field.

Another promising approach is the utilization of the domain structure. The highly symmetric crystal structures like cubic, tetragonal, and hexagonal lattices can host several crystallographically-equivalent axes, and thus form multiple domains with different direction of order parameters (such as P). If each domain is characterized by both magnetic and dielectric order parameters, ME effects are obtained from modulation of domain distribution by electric field (E) or magnetic field (H). Since these domains are energetically degenerated under zero external field, the H- (or E-) induced rearrangement of P-domain distribution can be preserved even after the removal of external field due to the existence of finite potential barrier. This strategy was first demonstrated on a ferroelectric weak ferromagnet (FM) $Ni_3B_7O_{13}I$[3] with persistent 180°-reversal of P-vector under 90°-rotation of H [26]. However, such ME control using domain structure has seldom been achieved, because of the rareness of similar ferroelectric ferromagnetic compounds [27] and the strict symmetry restrictions for selective domain switching [28]. The application of this approach to magnetically-induced ferroelectrics are also quite rare.

To apply multiferroics to the novel high-density non-volatile storage device, such as an H-controlled FE memory, the persistent switch of P-direction as well as increase of the number of switchable meta-stable states are highly desirable [29]. While several approaches including domain control have realized the H-induced persistent change of P-direction [26, 30, 31], they are all limited to the 180° switch between $\pm P$ states. Aside from the non-volatile nature, H-induced discontinuous switch of P-direction by other than 90° or 180° has never been achieved. In the following, we extend the concept of domain switching to demonstrate the persistent magnetic control of six ferroelectric domains with different P-directions for triangular lattice ferroelectric helimagnet $CuFe_{1-x}Ga_xO_2$ ($x = 0.035$). The flop of magnetic modulation vector is induced upon every 60°-rotation of in-plane H around

[3] In $Ni_3B_7O_{13}$, ferroelectricity and weak ferromagnetism arise from different origins.

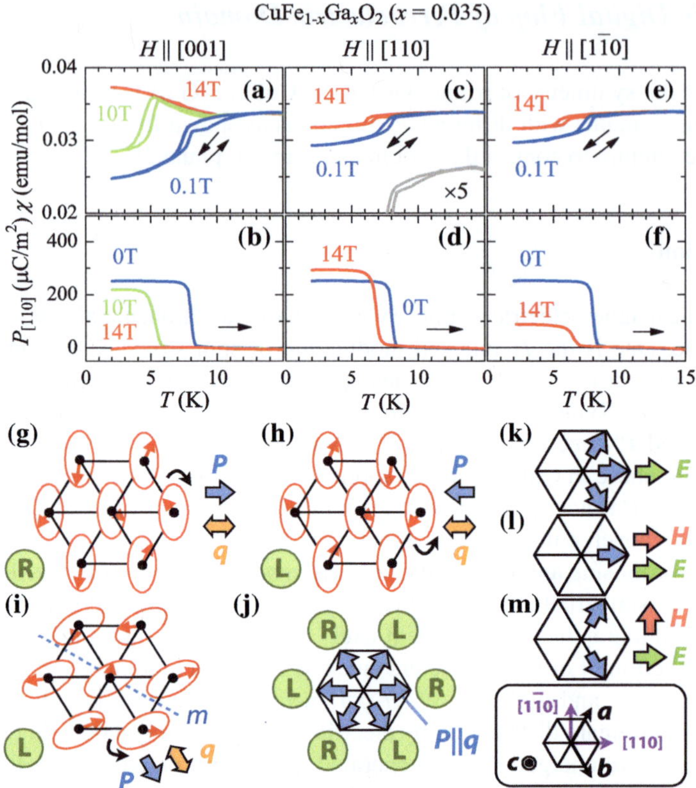

Fig. 3.14 Temperature (T) dependence of magnetic susceptibility (χ) and [110] component of electric polarization (P). Magnetic field (H) is applied along **a**, **b** [001], **c**, **d** [110], and **e**, **f** [1$\bar{1}$0], respectively. *Arrows* indicate the direction of T-scan. In **c**, a magnified profile of χ at 0.1 T (arbitrarily off-set) is also shown. **g–i** Three out of six possible multiferroic domains with proper screw magnetic structure on triangular lattice. *Circled* "R" and "L" denote the chirality of spin–spiral. *Directions* of P and magnetic q-vector are also indicated. **j–m** Distribution of multiferroic domain(s) favored under various H and electric field (E). The spin chirality corresponding to each P-domain is shown in **j**

the c-axis, which leads to every 120°-flop of P-vector within the triangular-lattice basal plane. Interestingly, the chirality of spin–spiral is always conserved upon the P-flop, which is discussed in the light of the stable structure of multiferroic domain wall (DW).

Results

In the ferroelectric proper screw (NC) magnetic phase of $CuFe_{1-x}Ga_xO_2$ (x = 0.035), $P \parallel q \parallel [110]$ relationship always holds (Fig. 3.14g)[4] and reversal of spin-chirality provides opposite direction of P-vector (Fig. 3.14h), as already mentioned in "Introduction". Without external field, this compound has six equivalent ferroelectric domains with $P \parallel \langle 110 \rangle$ (Fig. 3.14j) due to high symmetry of triangular lattice. These six ferroelectric domains possess one-by-one correspondence to six helimagnetic domains with three different $q \parallel \langle 110 \rangle$ and two spin-chiral degrees of freedom.

Since so far only the properties under zero magnetic field have been reported for Ga-doped $CuFeO_2$ [18], we first establish $H-T$ phase diagrams for the $CuFe_{1-x}Ga_xO_2$ (x = 0.035) specimen in $H_{[001]}$, $H_{[110]}$ and $H_{[1\bar{1}0]}$ (Fig. 3.15a–c). They are determined from the measurements of T- and H-dependence of M and [110] component of P ($P_{[110]}$) (Fig. 3.14a–f and Fig. 3.15d–f), by analogy with the case for $CuFe_{1-x}Al_xO_2$ (x=0.02) under $H_{[001]}$. Here, paramagnetic and two different sinusoidally-modulated collinear incommensurate magnetic phases [12] are referred to as PM, ICM1 and ICM2, respectively. With any direction of H, the boundary of the FE phase (shadowed region) always coincides with that of the NC magnetic phase, which ensures the coupling between ferroelectricity and proper-screw magnetic structure. In T-scan profiles, the onset of spiral magnetic order, coupled with the emergence of ferroelectric P, can be detected as the sudden drop of magnetic susceptibility $\chi(= M/H)$. While the NC(FE) phase is replaced by CM5 under $H_{[001]} > 12$ T, we found that NC(FE) survives against in-plane H ($H_{[110]}$ and $H_{[1\bar{1}0]}$) up to 14 T.

Even with in-plane H, the magnitude of $P_{[110]}$ shows significant H-dependence, as shown in Fig. 3.14d and f: $P_{[110]}$ increases with $H_{[110]}$ and decreases with $H_{[1\bar{1}0]}$. This behavior can be interpreted as the rearrangement of multiferroic domains, as described below. In general, antiferromagnetically ordered spin moments prefer to lie within the plane perpendicular to H, as typically observed for the spin flop transition. Thus, in case of proper screw magnetic structure, H favors the magnetic domain with $q \parallel H$. In contrast, electric field E affects the selection of spin chirality [8]. If E is applied along [110], three multiferroic domains are selected as depicted in Fig. 3.14k. Further application of H should sort domains; a domain distribution as illustrated in Fig. 3.14l or m is favored with $H_{[110]}$ or $H_{[1\bar{1}0]}$. Comparing these arrangements (Fig. 3.14k–m), $P_{[110]}$ should increase with $H_{[110]}$ and decrease with $H_{[1\bar{1}0]}$. This idea well explains the observed experimental results. With $H_{[110]}$- and $H_{[1\bar{1}0]}$-scan at 2 K, both M and $P_{[110]}$ show anomalies around 5 T only in the field-increasing run (Fig. 3.15e and f), suggesting the field-irreversible rearrangement of domain distribution at this specific field direction.

Next, we have investigated the vector components of P in response to the in-plane H rotating around the c-axis. For this purpose, the both [110] and [1$\bar{1}$0] components of P were measured simultaneously with two pairs of electrodes. In this configuration, both P and H can be expressed as the two-dimensional vector on the (001) plane. Hereafter, we define the angle between H (P) and the [110]-axis as θ_H (θ_P).

[4] Here, we define the magnetic q-vector as the director with no $+/-$ sign.

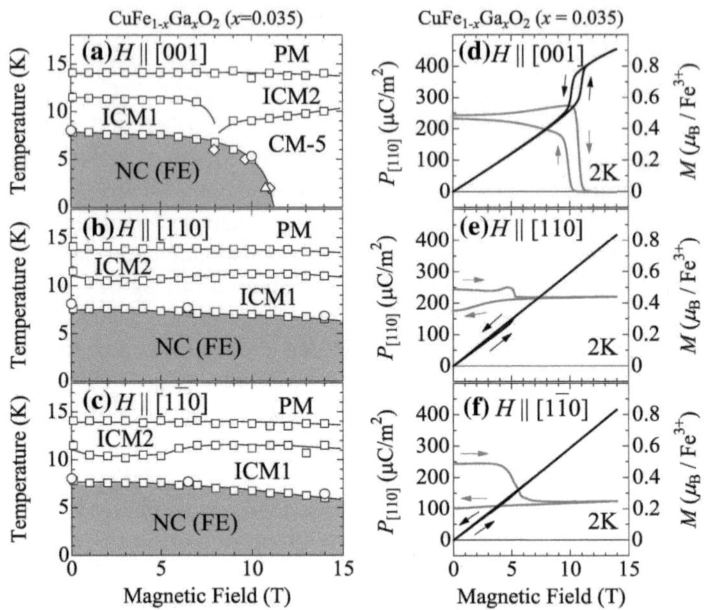

Fig. 3.15 $H-T$ phase diagrams with H parallel to **a** [001], **b** [110], and **c** [1$\bar{1}$0] direction. *Circles*, *triangles*, *squares*, and *diamonds* are the data points obtained from $P-T$, $P-H$, $\chi-T$, and $M-H$ curves, respectively. All the data were taken from the increasing T or H runs. Ferroelectric (FE) state is observed in the shadowed region. **d–f** H-dependence of P (*gray line*) parallel to [110] and M (*black line*), with H applied along **d** [001], **e** [110], and **f** [1$\bar{1}$0] direction. H was swept from 0 to 14 T and then back to 0 T, after T was lowered to 2 K at 0 T. The *arrows* indicate the direction of H-scan

Since the specimen was cooled with H and E both applied along [110], we assume the uniform initial domain state as shown in Fig. 3.14l.

Figure 3.16a and d shows $P_{[110]}$ and $P_{[1\bar{1}0]}$ as a function of θ_H, measured at $H = 6.5$ T without E. Both $P_{[110]}$ and $P_{[1\bar{1}0]}$ show a periodic change with the cycle of 180°. To see the development of P more directly, we plot the obtained θ_P against θ_H for the θ_H-increasing run (Fig. 3.17a). In agreement with the expected initial state in Fig. 3.14l, the relationship $P \parallel H \parallel [110]$ is confirmed at $\theta_H = 0$. As θ_H increases, P suddenly flops by about 120° at $\theta_H = 41°$, and the relation that $-P \parallel H \parallel [100]$ holds at $\theta_H = 60°$. Since H favors domains with $q \parallel P \parallel \pm H$, this transition can be considered as the flop of q-vector from $q \parallel [110]$ to $q \parallel [100]$. Note that both transitions to $P \parallel [100]$ and $P \parallel -[100]$ states seem possible from the $P \parallel [110]$ initial state, but in reality only the $P \parallel -[100]$ state is selected here. Such a q-flop (or 120°-flop of P) is observed on every 60°-rotation of H, consistent with the symmetry of underlying triangular lattice. The absolute value $|P|$ (not shown) is almost constant for any θ_H.

Between the two opposite directions of H-rotation, a relatively large hysteresis is found in P. If we estimate the critical θ_H from the average of both direction of scans,

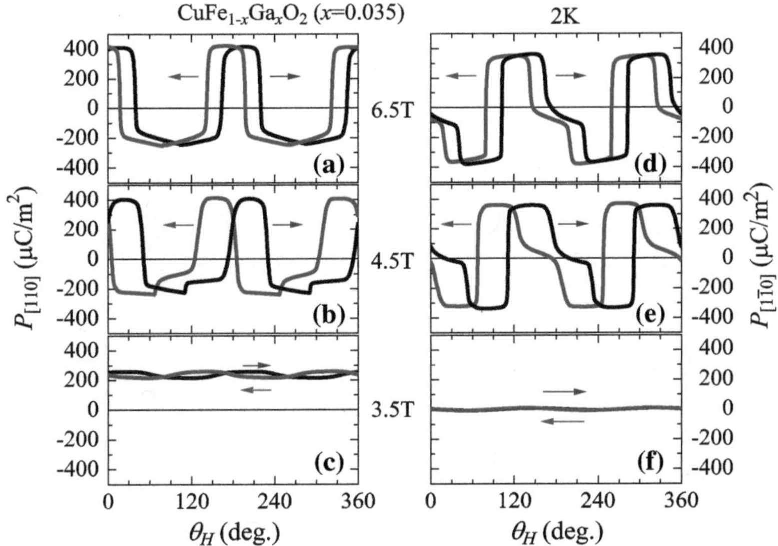

Fig. 3.16 **a–c** [110] and **d–f** [$1\bar{1}0$] components of P simultaneously measured in H rotating within the (001) plane. θ_H denotes the angle between H-vector and the [110] axis (see Fig. 3.17d). *Arrows* indicate the direction of H-rotation. Absolute value of P was determined by T-scan

the q-flop is always centered at $\theta_H = (30 + 60n)^\circ$ (n: integer). This agrees with the equilibrium point of two magnetic q-vectors. The appearance of hysteresis means that an excess gain of Zeeman energy is needed to overcome the potential barrier height. When H is reduced from 6.5 to 4.5 T, the reduction of Zeeman energy leads to expansion of the hysteresis (Fig. 3.16b and e). Below 3.5 T, the potential barrier cannot be overcome and no q-flop behavior is observed (Fig. 3.16c and f).

Corresponding variation of M as a function of θ_H is shown in Fig. 3.17b and c. To obtain the $q \parallel [110]$ initial state, the specimen was cooled at $\theta_H = 0^\circ$ with $H = 6.5$ T and then magnitude of H is fixed prior to measurements. Without q-flop, this initial state should give $\Delta M \propto (\chi_\parallel - \chi_\perp)\cos(2\theta_H)$, where $\chi_\parallel$ and $\chi_\parallel$ denotes χ parallel or perpendicular to q. This agrees well with θ_H dependence of M at 3 T (Fig. 3.17c), indicating the robustness of single-q state and the non-volatile nature of q-domain distribution. In contrast, the profile at 6.5 T has a period of 60° and corresponding sinusoidal curve is shifted by 60° upon every P-flop transition. This confirms the emergence of q-flop as the origin of P-flop and the clamping of ferroelectric and magnetic DWs.

Figure 3.17d illustrates the relationship between P and H at θ_H=$(60n)^\circ$. At each θ_H, we could confirm $P \parallel \pm H$, in agreement with the q-flop model. Importantly, upon each transition, $P \parallel H$ and $P \parallel -H$ alternately appears. Generally, magnetic domains can be mutually converted by symmetry operation that is broken by magnetic order [28]. If we apply space inversion to Fig. 3.14g, a domain with opposite P and reversed spin chirality can be obtained (Fig. 3.14h). Mirror operation on Fig. 3.14g

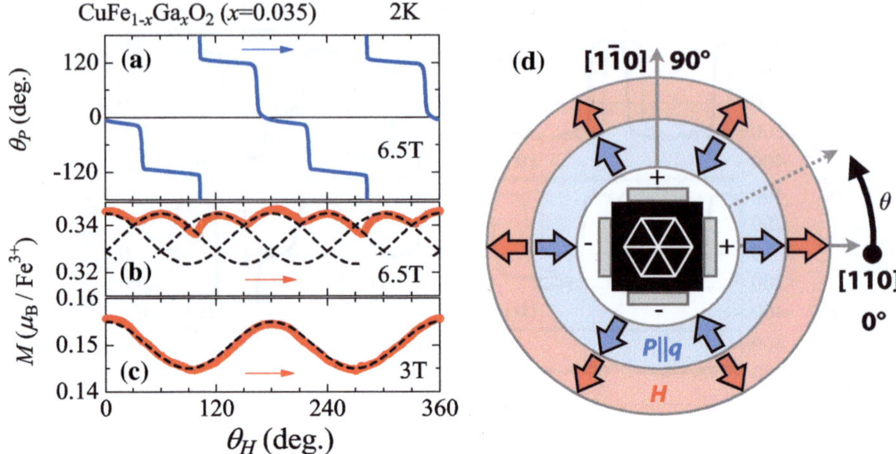

Fig. 3.17 **a** Relationship between the directions of P and H, both confined within the (001) plane. θ_P (θ_H) denotes the angle between P- (H-) direction and the [110] axis. **b**, **c** Corresponding variations of M. *Dashed lines* indicate theoretically expected behaviors. **d** Relationship among P, q and H at $\theta_H = (60n)^\circ$ (n: integer)

generates another domain with reversed spin chirality (Fig. 3.14i). Likewise, we can reproduce all six P-domains and determine their corresponding spin chirality (Fig. 3.14j). From this relationship, it is concluded that the chirality of spin–spiral is always conserved upon the q-flop.

Discussion

As demonstrated above, the employment of highly-symmetric crystal structure offers multiple multiferroic domains, which enables H-induced non-volatile switching of P-vector among as many as six possible directions.

Here, the problem is the reason why the spin-chirality is always preserved upon the q-flop transition. Since two spin-chiral domains are allowed for the selected q-vector and they are energetically degenerated under applied magnetic field, selection of odd chirality upon q-flop should reflect the energy difference between two possible domain wall structures connecting domains with the same or opposite chirality. Notably, without the contribution from domain wall, equal population of opposite spin-chiral domains appear upon the q-flop and hence give $P = 0$. Thus, the nature of multiferroic domain wall plays an crucial role to enable the H-induced selective P-domain switching.

At this stage, the physics of domain wall in ferroelectric helimagnets is still yet to be established. Recently, Kagawa et al. calculated the stable domain wall structure for ferroelectric cycloidal magnet $DyMnO_3$, which connects two domains with different spin–spiral plane but with the same q-vector [32]. In this case, the cycloidal spin

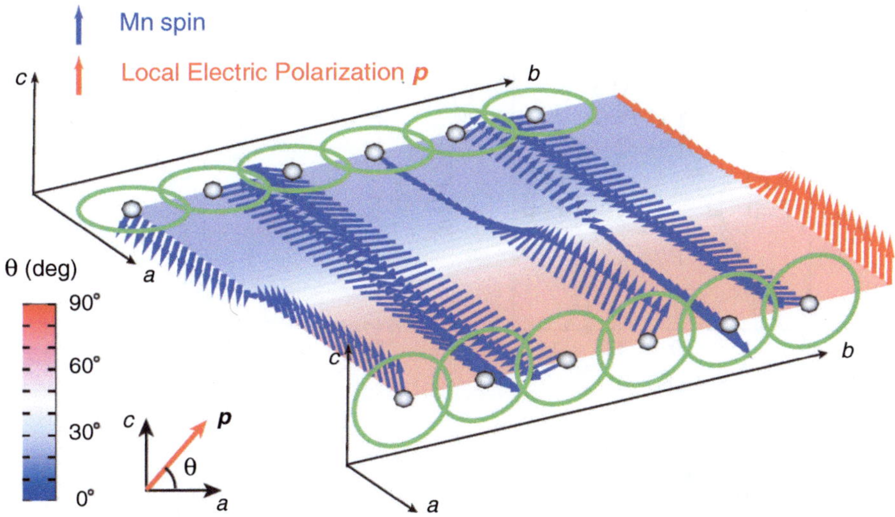

Fig. 3.18 Calculated domain wall structure between the $P \parallel +c$ (*bc*-cycloidal) and $P \parallel +a$ (*ab*-cycloidal) domains for 36 Mn sites, which assumes the case for $DyMnO_3$. Here, the magnetic q-vector is fixed along the b-axis. *Blue* and *red arrows* represent the Mn spins and the local electric polarizations, respectively. The color gradation represents the angle of local electric polarization relative to the a-axis. The angle becomes 45° along the domain wall center, which runs parallel to the b-axis. (Adapted with permission from [32], ©2009 APS.)

plane and associated P-direction continuously rotate within the domain wall region (Fig. 3.18).[5] The calculated thickness of this multiferroic domain wall is about 20 unit cells, while the domain wall in conventional ferroelectrics is generally atomically thin. This probably reflects the magnetic origin of ferroelectricity and weakness of induced P-value. Here, the thickness of multiferroic domain wall is governed by the ratio between magnetic anisotropy and exchange coupling.

Interestingly, this type of multiferroic domain wall also plays an essential role in the H-induced P-flop transition in $MnWO_4$ [33]. In this compound, application of unidirectional $H \parallel b$ induces the transition from bc-spiral magnetic state ($P \parallel b$) to ac-spiral magnetic state ($P \parallel a$), leaving the $q \parallel c$ unchanged. While two spin-chiral states ($P \parallel \pm a$) are energetically degenerated under applied magnetic field, slight tilt of H-direction from the b-axis lifts the degeneracy of two possible domain walls connecting ($P \parallel +b \rightarrow P \parallel +a$) or ($P \parallel +b \rightarrow P \parallel -a$). Thus, the sign of electric polarization after the spin-flop transition can be determined by the tilting direction of H from the b-axis (Fig. 3.19). When H is precisely applied along the b-axis, equal population of $P \parallel \pm a$ domains (i.e. $P = 0$) appear. Similar phenomena has also been reported to $TbMnO_3$ [34].

[5] This multiferroic domain wall shows dielectric relaxation behavior in the frequency range of $1 \sim 0.1$ MHz, which turns out to be the origin of giant magnetocapacitance effect observed in $DyMnO_3$ [32].

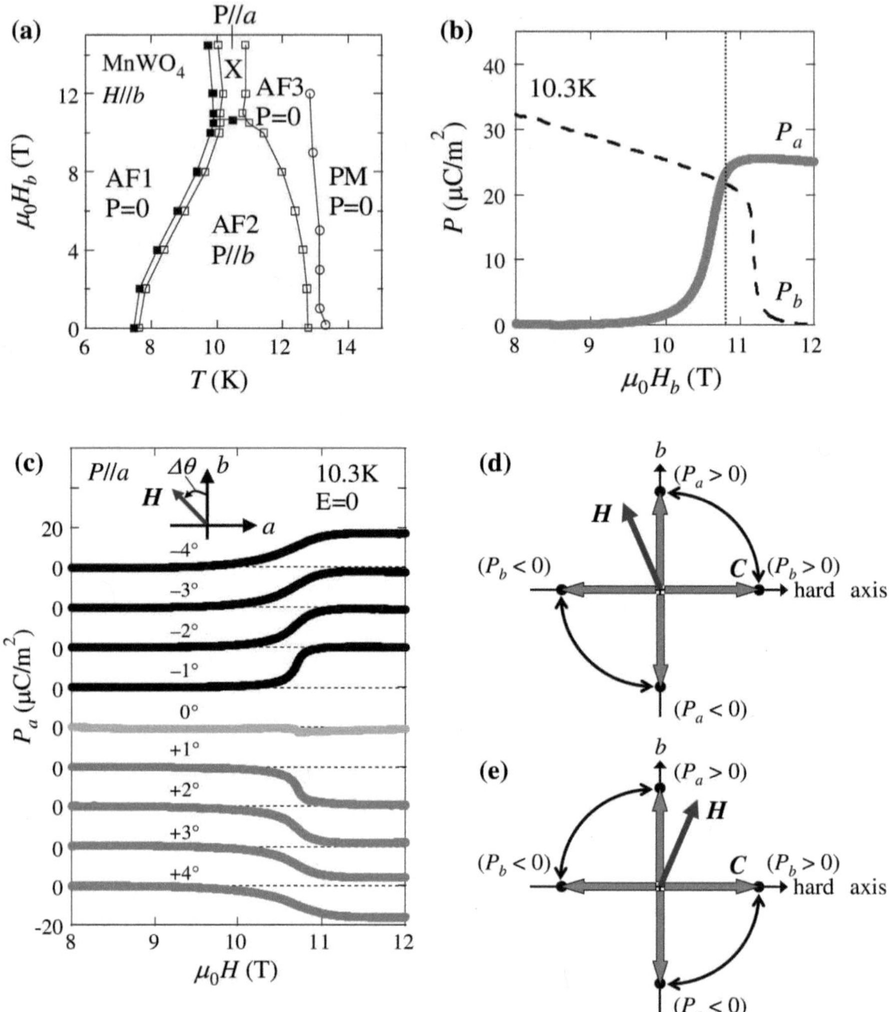

Fig. 3.19 **a** Magnetoelectric phase diagram of $MnWO_4$ in magnetic field parallel to the b-axis (H_b). **b** H_b-dependence electric polarization P. **c** H_b-dependence of P_a measured at 10.3 K, with H slightly tilted from the b-axis toward the direction of the a-axis by angle of $\Delta\theta$. The measurement was performed in the H-increasing run from the negatively-poled $P \parallel b$ state at 8 T. **d** and **e** Favorable rotation pattern of the vector spin chirality ($\mathbf{C} = \mathbf{S}_i \times \mathbf{S}_j$) under a canted H. The P-direction corresponding to each **C** are also indicated. (Adapted with permission from [33], ©2008 APS.)

In contrast, the situation seems to be more difficult in case of the q-flop transition. Murakawa et al. has recently investigated the q-flop behavior in a ferroelectric helimagnet $ZnCr_2Se_4$ with cubic lattice. Upon 90°-flop of q-vector under rotating H on $ZnCr_2Se_4$, the spin chirality is preserved in lower-H region but reversed in

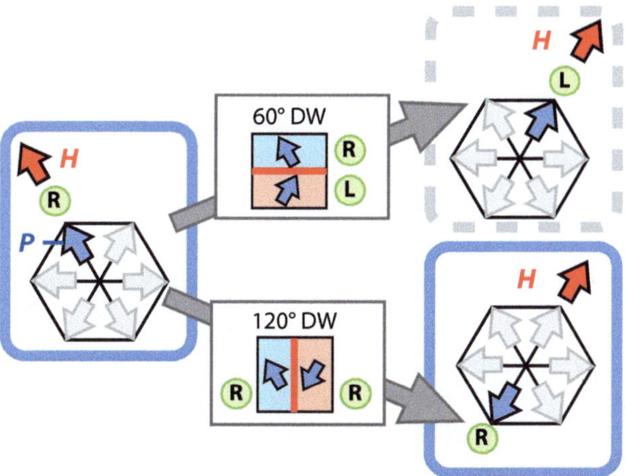

Fig. 3.20 Two possible P-domains after the H-induced q-flop transition in $CuFe_{1-x}Ga_xO_2$ ($x = 0.035$). While the two spin-chiral domains for the selected q-vector are energetically degenerated under magnetic field, the energy difference between two possible domain walls determines the spin-chirality after the q-flop transition. Here, *circled* "R" and "L" denote the chirality of spin–spiral. *Blue* and *red arrows* indicate the direction of P and H, respectively

higher-H region [35]. This implies that the relative stability of multiferroic domain wall depends on the magnitude of applied magnetic field. In the present case of $CuFe_{1-x}Ga_xO_2$, when rotating H induces the P-flop from P_1 to P_2, the angle between P_1 and P_2 across the DW becomes 120° for the same chirality (120°-DW), and 60° for the opposite chirality (60°-DW). The experimentally observed robustness of spin chirality can be explained, only provided that 120°-DW is more stable than the 60°-DW (Fig. 3.20). So far, the microscopic structure of multiferroic domain wall connecting different magnetic q-vectors is not obvious. Detailed theoretical calculation of relative stability of two types of DWs considering all magnetic, dielectric, structural, and chiral degree of freedoms, as well as direct observation of domain wall structure, is highly desirable.

Conclusion

In this section, we demonstrate the persistent magnetic control of six P-domains for triangular-lattice helimagnet $CuFe_{1-x}Ga_xO_2$ ($x = 0.035$). The flop of the magnetic q-vector is induced by every 60°-rotation of in-plane H around the c-axis, which leads to every 120°-flop of P-direction within the triangular-lattice basal plane. The chirality of spin–spiral is always conserved upon the P-flop, which may reflect the stability of the specific multiferroic domain wall structure. In more general, the nature of multiferroic domain wall in a ferroelectric helimagnet should play a key role in determining the P-direction or spin-chirality upon the H-induced q-flop; this may enable the unusual selective P-domain switching with varying direction and magnitude of H. Ferroelectric helimagnets with highly symmetric crystal lattice like cubic, tetragonal or hexagonal lattices can host multiple H-switchable P-domains as revealed here, and hence are promising for magnetoelectric control with non-volatility and multiple-valued nature.

Electromagnon in the Paraelectric Collinear Spin State

In the following, we investigate the dynamical aspects of $CuFe_{1-x}Ga_xO_2$. Strong magnetoelectric coupling in magnetically-induced ferroelectrics often provides a novel collective excitation called "electromagnon" (i.e. magnon driven by a.c. electric field) in the dynamical regime. For the search of such unique excitation in $CuFe_{1-x}Ga_xO_2$, we employed the THz time-domain spectroscopy.

Introduction

In the static regime, so far at least two microscopic ME coupling mechanisms have been established: Local electric polarization $\vec{P}_{ij}$ produced between two magnetic sites is described as

$$\mathbf{P}_{ij} = \mathbf{\Pi}_{ij}(\mathbf{S}_i \cdot \mathbf{S}_j) + A\mathbf{e}_{ij} \times (\mathbf{S}_i \times \mathbf{S}_j), \tag{3.2}$$

where $\mathbf{\Pi}_{ij}$ is a vector unique to the underlying crystal structure, A a coupling coefficient, and $\mathbf{e}_{ij}$ an unit vector connecting two magnetic moments $\mathbf{S}_i$ and $\mathbf{S}_j$, respectively. The first term represents the exchange striction, and the second term comes from the inverse effect of Dzyaloshinskii–Moriya (DM) interaction [36]. Here, the inverse DM mechanism relies on the spin–orbit interaction, but the exchange striction does not. Furthermore, some other ME coupling mechanism originating from spin–orbit interaction has also been suggested for $CuFe_{1-x}Ga_xO_2$ [9, 10]. With any mechanism, a modification of magnetic structure leads to a significant change of induced electric polarization P.

One important consequence of such a strong ME coupling is the appearance of a novel collective excitation called electromagnon (i.e. magnon driven by a.c. electric field E^{ω}) in the dynamical regime. With detailed polarization analyses of absorption spectra, existence of electromagnon (EM) excitation has been established for ferroelectric (FE) helimagnets RMnO$_3$ [37, 38], RMn$_2$O$_5$ [39], and $Ba_2Mg_2Fe_{12}O_{22}$ [40, 41]. Here, the most crucial is the microscopic origin of dynamical ME coupling, which is not necessarily identical to that of the magnetically-induced static P in the same compound. According to the inverse DM scheme, the EM excitation in FE helimagnets emerges as the rotational oscillation of spin–spiral plane and associated P-vector [42]. While this rotational mode should be active only with E^{ω} perpendicular to the spin–spiral plane, the observed selection rule for RMn$_2$O$_5$ contradicted with this prediction [39, 43]. For $RMnO_3$ and $Ba_2Mg_2Fe_{12}O_{22}$, the selection rules remain unchanged even after the spin-flop transition under applied H [38, 40]. Latest theoretical studies suggested that the exchange striction mechanism can also host the EM activity in noncollinear magnets, but with the selection rule tied to the chemical lattice [44, 45]. Importantly, this exchange-striction-induced EM is inactive in the collinear magnetic phase, since the differential polarization $\delta P_{ij} \propto S_i \cdot \delta S_j$ becomes always zero ($S_i \perp \delta S_j$). This latter model well reproduces the observed selection rules or absorption spectra of EM in all the three helimagnetic compounds [40, 43–46], whereas the firm experimental evidence of spin–orbit coupling mediated EM excitation is still lacking.

In the following, we report the experimental discovery of electromagnon excitation in the paraelectric collinear magnetic ($\uparrow\uparrow\downarrow\downarrow$) phase of triangular lattice antiferromagnets $CuFe_{1-x}Ga_xO_2$. This EM mode was found to rather vanish in the FE helimagnetic phase. The anti-correlation between the electromagnon and the noncollinear magnetism excludes the exchange-striction mechanism as the origin of dynamical ME coupling, and hence suggests that the observed electromagnon is electrically activated by the spin–orbit coupling.

Result

The detailed experimental setup for THz time-domain spectroscopy is described in "THz Time-Domain Spectroscopy". By using this experimental methods, we can obtain the spectrum of complex transmittance t. It is further converted into complex refractive index $n = \sqrt{\epsilon\mu}$ using the following relationship;

$$t = \frac{2\mu}{n+\mu}\frac{2n}{n+\mu}\exp\Big[-i\frac{\omega}{c}d(n-1)\Big], \tag{3.3}$$

where ϵ, μ, d, ω, and c represent the complex dielectric constant, complex magnetic permeability, sample thickness, frequency of light, and velocity of light, respectively. To numerically solve Eq. (3.3), we approximate the pre-exponential factor by $4n/(n+1)^2$ assuming $\mu \simeq 1$ unless otherwise noted. As shown later, this approximation hardly affects the obtained n (or $\epsilon\mu$) spectrum.

We first investigated the low-energy electrodynamics in the paraelectric CM4 collinear magnetic phase (Fig. 3.21e).[6] Figure 3.22a and b indicates the real and imaginary part of $\epsilon\mu$ spectra (Re[$\epsilon\mu$] and Im[$\epsilon\mu$]) with various polarization configurations for the $CuFe_{1-x}Ga_xO_2$ ($x = 0.01$) specimen at 4.4 K, respectively. With $E^\omega \parallel [110]$ and $H^\omega \parallel [1\bar{1}0]$, two resonance modes are observed at 1.2 and 2.3 meV. Only the former one survives for $E^\omega \parallel [001]$ and $H^\omega \parallel [1\bar{1}0]$, whereas only the latter one does for $E^\omega \parallel [110]$ and $H^\omega \parallel [001]$. These results unveil that the excitation

[6] Symmetry of triangular lattice allows the existence of three equivalent $\vec{q} \parallel \langle 110\rangle$. The presently observed spectra reflect the contributions from all the three q-domains.

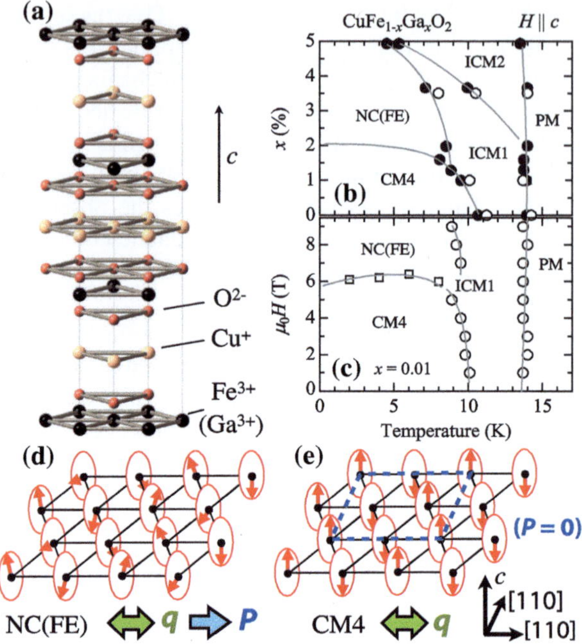

Fig. 3.21 **a** Crystal structure and **b** $x-T$ magnetic phase diagram at zero magnetic field ($H = 0$) for $CuFe_{1-x}Ga_xO_2$. **c** $H-T$ magnetic phase diagram for the $x = 0.01$ specimen with static H applied parallel to the c-axis. *Circles* and *squares* are the data points obtained from the measurements of magnetization with T- and H-increasing runs, respectively. *Open* (*closed*) *symbols* represent the data points determined in the present work (the previous work by Terada et al. [16]). **d** and **e** indicate magnetic structures of the NC(FE) and CM4 phase, respectively. *Dashed square* in **e** represents the magnetic unit cell in the CM4 phase. The directions of magnetic q-vector and electric polarization P are also indicated

at 2.3 meV is an EM mode driven by $E^{\omega} \parallel [110]$, while the one at 1.2 meV is a conventional magnon mode driven by $H^{\omega} \parallel [1\bar{1}0]$.

In the following, we focus on the behavior with the $E^{\omega} \parallel [110]$ and $H^{\omega} \parallel [1\bar{1}0]$ configuration. To further analyze the aforementioned $\epsilon\mu$ spectrum, the corresponding absorption coefficient $\alpha(= -2(\ln|t|)/d)$ and the decomposed ϵ and μ spectra are plotted in Fig. 3.23a–c, respectively. To discriminate the ϵ- and μ-contributions to the $\epsilon\mu$ spectrum, we first assumed $\mu = 1$ for $\hbar\omega > 2.0$ meV. The obtained ϵ spectrum can be fitted well with the sum of two Lorentzian functions; a higher-frequency mode represents the lowest-lying optical phonon to give rise to the tail absorption observed below 5 meV. By substituting this ϵ-fitting function into Eq. (3.3), the μ spectrum is deduced for $\hbar\omega < 2.0$ meV. It can be fitted as well with a single Lorentzian function. We find that the relationship $|\mu - 1| < 0.1$ always holds, which justifies

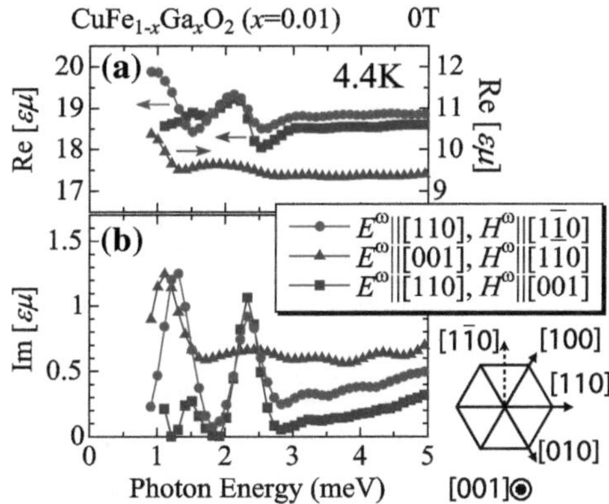

Fig. 3.22 Real and imaginary parts of $\epsilon\mu$ spectra (Re [$\epsilon\mu$] and Im [$\epsilon\mu$]) for the $x = 0.01$ specimen measured at 4.4 K with various light-polarization configurations

the aforementioned $\mu \simeq 1$ approximation adopted for the pre-exponential factor in Eq. (3.3).[7]

In Fig. 3.24b, we show T-dependence of Im[$\epsilon\mu$] spectrum for the $x = 0.01$ specimen. Increase of temperature leads to broadening of two resonance peaks, and they become almost undiscernible in the ICM1 magnetic phase above 10 K. ICM1 and ICM2 are partially disordered magnetic phases [5, 7, 12], and may lose the spin correlation enough for magnons or EM excitations to be observed. The undoped $x = 0.00$ specimen also has two resonance modes at the same frequency in the CM4 magnetic ground state (Fig. 3.24a), and shows similar T-dependence of Im[$\epsilon\mu$] spectrum as observed for the $x = 0.01$ specimen. In contrast, the $x = 0.035$ specimen with the ferroelectric NC helimagnetic ground state (Fig. 3.21d) shows no discernible peak structure in the whole temperature range (Fig. 3.24c).[8] To summarize, the EM excitation driven by $E^{\omega} \parallel [110]$ is active only in the paraelectric collinear CM4 magnetic phase, not in the ferroelectric NC helimagnetic phase.

Next, we discuss the microscopic origin of these excitations. The spin-wave (SW) dispersion for $CuFeO_2$ has been investigated by a previous inelastic neutron diffraction study [20], and the analysis clarified the existence of two SW branches as reproduced in Fig. 3.23g [21]. In general, an antiferromagnetic resonance (AFMR) appears as the excitation of zone center mode at $\vec{k} = 0$ by H^{ω} perpendicular to the collinear spin direction (Fig. 3.23e). From this criterion, we concluded the excitation at 1.2 meV driven by $H^{\omega} \parallel [1\bar{1}0]$ is AFMR on the lower SW branch. In contrast, the excitation energy of the observed EM (~ 2.3 meV) agrees with that of the zone-center mode on the upper SW branch.

So far, the most successful scheme to explain the dynamical ME coupling is the exchange striction. However, this mechanism is inactive in the collinear spin system

[7] To further check the validity of this approximation, we deduced $\epsilon\mu$ spectrum assuming $\mu = 1.1 + 0.1i$ for the pre-exponential factor in Eq. (3.3). The result coincides with the one calculated with the original $\mu = 1$ assumption within the experimental error.

[8] We also measured the Im[$\epsilon\mu$] spectrum for the $x = 0.035$ specimen under the $E \parallel [001]$ and $H \parallel [1\bar{1}0]$ condition, but no peak structure could be observed.

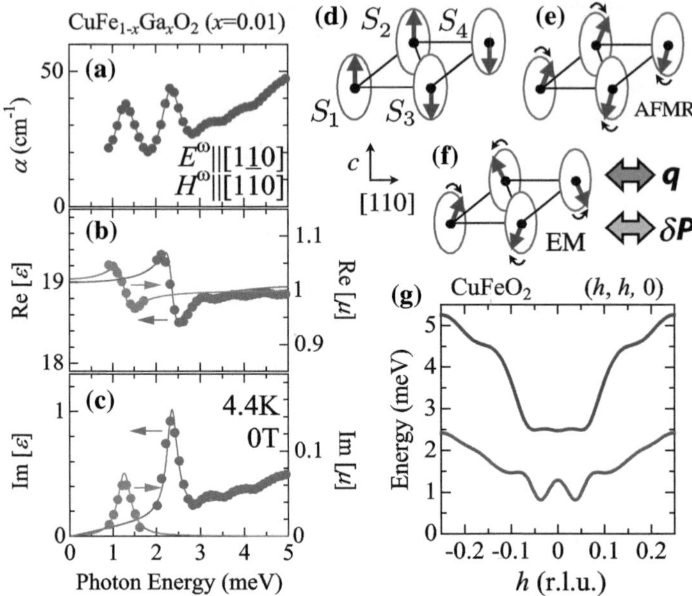

Fig. 3.23 **a–c** Absorption coefficient α, real and imaginary part of ϵ and μ spectra for the $x = 0.01$ specimen measured at 4.4 K with $E^{\omega} \parallel [110]$ and $H^{\omega} \parallel [1\bar{1}0]$. *Solid lines* in **b** and **c** represent the fits with the sum of Lorentzian functions. **d** Spin structure of the CM4 magnetic ground state. **e** and **f** indicate the possible excitation modes corresponding to the observed genuine magnon (AFMR) and electromagnon (EM), respectively. In (f), S_1 and S_3 rotate to the opposite direction of S_2 and S_4 within a plane perpendicular to $\vec{q}$. **g** Spin-wave (magnon) dispersion of $CuFeO_2$ along the $(h, h, 0)$ direction as proposed by Fishman et al. [21]

like the present CM4 phase, since the relationship $\delta P_{ij} \propto S_i \cdot \delta S_j = 0$ always holds [44, 45]. This strongly suggests the relevance of spin–orbit coupling to the present EM mode; the relatively weak peak intensity in the Im[ϵ] spectrum (one order of magnitude smaller than that of $DyMnO_3$ [38]) also supports this scenario. Considering that the static P in NC(FE) is induced by the proper screw magnetic order through the spin–orbit interaction mediated modulation of Fe $3d-O2p$ hybridization [10, 9], we may anticipate the analogous origin for the presently observed dynamical ME coupling in CM4. For example, the magnetic excitation as depicted in Fig. 3.23f can dynamically generate a proper-screw-like spin texture with a finite spin chirality, which is expected to induce non-zero electric dipole along the $\delta P \parallel q \parallel [110]$ direction. This mode should be active only with $E^{\omega} \parallel [110]$, which is consistent with the experimental results. The disappearance of EM mode in the NC(FE) phase may reflect the alteration of magnetic symmetry or Brillouin-zone folding, but the detail is left to be clarified.

We further investigated the development of the EM mode in static H applied along the [001] direction. Figure 3.24d indicates the H-dependence of Im[ϵ] spectrum measured at 6 K for the $x = 0.01$ specimen with $E^{\omega} \parallel [110]$ and $H^{\omega} \parallel [001]$, where

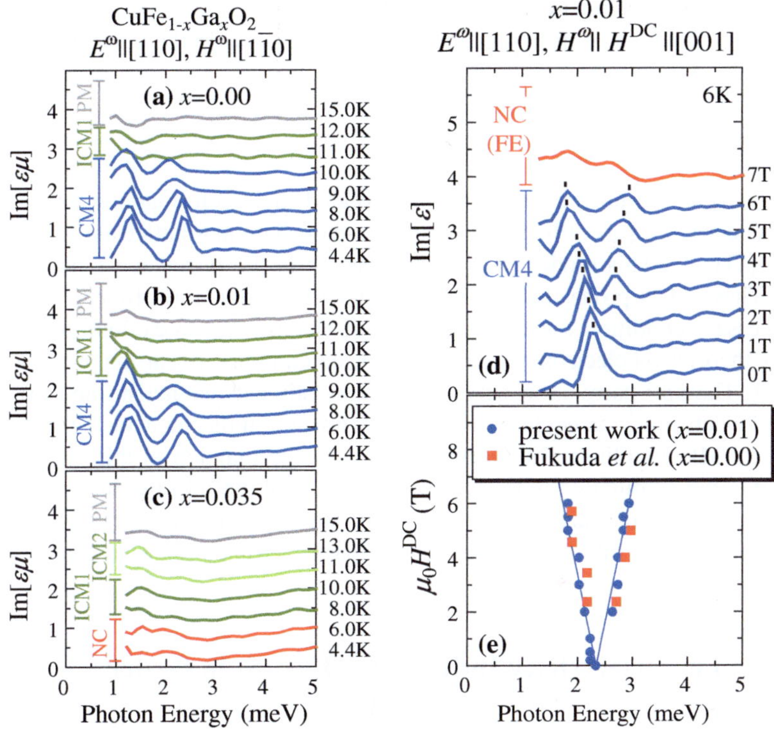

Fig. 3.24 Temperature dependence of Im[$\epsilon\mu$] spectra measured with $E^\omega \parallel [110]$ and $H^\omega \parallel [1\bar{1}0]$ for **a** $x = 0.00$, **b** $x = 0.01$, and **c** $x = 0.035$ specimens, respectively. **d** Im[ϵ] spectra for the $x = 0.01$ specimen measured at 6 K with $E^\omega \parallel [110]$ and $H^\omega \parallel [001]$ in various magnitudes of static H (H^{DC}) applied along the [001] direction. H-dependence of observed EM peak positions (*circle*), as well as the development of resonance modes previously reported by the ESR study for the $x = 0.00$ specimen [47] (*square*), are plotted in **e**

only the EM excitation can be observed. As H increases, the EM mode is found to split linearly in terms of H and form two peak structures. This reflects the H-linear splitting of SW branches, which is generally expected in collinear antiferromagnets with H applied parallel to the magnetic easy axis. Since the spectral shape of the two-magnon excitation should be independent of the external H [48], this ensures that the present electromagnon is excited by the one-magnon process. The observed evolution of EM peak positions under applied H is summarized in Fig. 3.24e. Note that similar H-dependence of resonance modes has been reported by Fukuda et al. from the ESR study for the $x = 0.00$ specimen [47], while they conventionally assigned these modes to AFMR driven by H^ω.[9] The peak structure observed in Im[ϵ]

[9] Our present results imply that the resonance modes found in the previous ESR study [47] is primarily driven by E^ω-component of incident microwave.

spectrum becomes almost invisible after the transition from the CM4 into NC(FE) phase at 6.3 T, which confirms the inactivity of EM mode in the latter NC phase.

Conclusion

In this section, we have experimentally revealed the electromagnon excitation in the paraelectric ↑↑↓↓ collinear magnetic phase of triangular lattice antiferromagnet $CuFe_{1-x}Ga_xO_2$. This mode was found to vanish in the ferroelectric helimagnetic phase. These facts prove that neither ferroelectricity nor noncollinear magnetism is a necessary condition for the appearance of electromagnon excitation, while the existing theories on electromagnon have focused on noncollinear magnets like helimagnets. The anti-correlation between the noncollinear magnetism and the emergence of electromagnon excludes the exchange striction mechanism as the origin of dynamical ME coupling: The electric activity of the magnon in this compound is ascribed to the modulation of the *p*-*d* hybridization at the spin-twisted excited state via the spin–orbit interaction. Our discovery suggests that similar electromagnon modes will be observable in a wide range of paraelectric collinear magnets.

ACrO$_2$: Ferroelectricity Induced by 120°-Spin Order

Introduction

With classical Heisenberg spins and dominance of nearest neighboring interaction, triangular lattice antiferromagnet generally favors the 120° spiral spin structure at the ground state. Here, each neighboring pair of magnetic moments makes an angle of 120°. Depending on the sign of anisotropy term $\mathcal{H}' = D\sum(S_i^z)^2$, the spin–spiral is confined in the plane parallel ($D > 0$: easy-plane type) or perpendicular ($D < 0$: easy-axis type) to the triangular lattice plane [1].

Notably, the inverse D–M model denies the appearance of ferroelectricity for such 120°-spin ordered states. Fig. 3.25a illustrate the easy-plane type 120°-spin order, and local P-direction along each chain expected from the inverse D–M model. Because of its threefold rotational symmetry, in-plane P component always cancels out. Similar cancelation is also predicted for the easy-axis type 120°-spin order. Nevertheless, recent discovery of ferroelectricity in $CuFe_{1-x}Ga_xO_2$ with proper screw spin order suggested that the inverse D–M mechanism is not the only magnetoelectric coupling mechanism to induce ferroelectricity on the triangular lattice. In the following, we investigate the magnetoelectric response of triangular lattice antiferromagnet ACrO$_2$, and demonstrate that the easy-axis type 120°-spin order can actually induce ferroelectricity. Combined the present results with the recent report for $RbFe(MoO_4)_2$ with the easy-plane anisotropy [49], we can predict that a broad range of trigonal

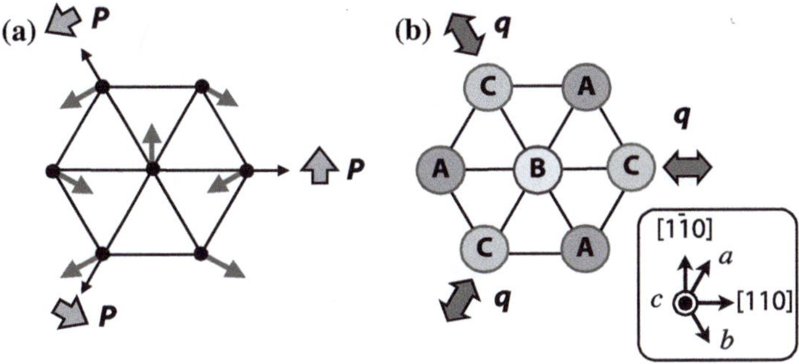

Fig. 3.25 **a** Schematic illustration of the easy-plane type 120°-spin order, where spin–spiral lies parallel to the triangular lattice basal plane. P-direction expected from the inverse D–M model along each chain are also indicated. **b** Magnetic modulation vectors (q) of 120°-spin order

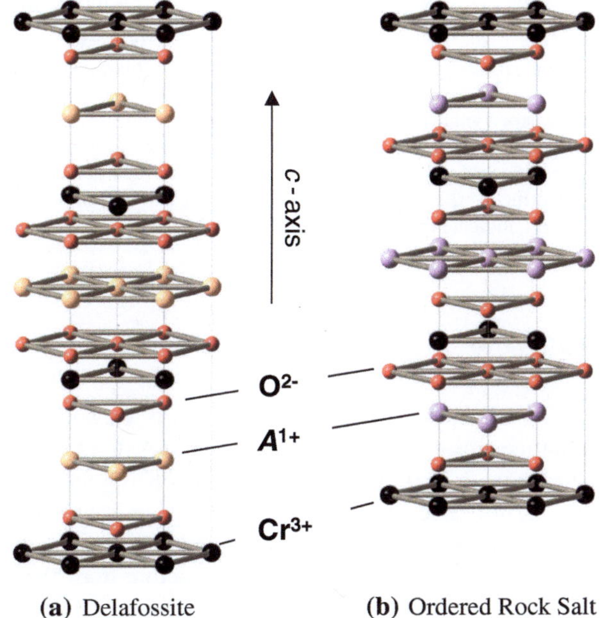

Fig. 3.26 Crystal structure of $ACrO_2$: **a** delafossite structure (A = Cu or Ag) and **b** ordered rock salt structure (A = Li or Na)

materials with the 120°-spin structure can be multiferroic, irrespective of their magnetic anisotropy.

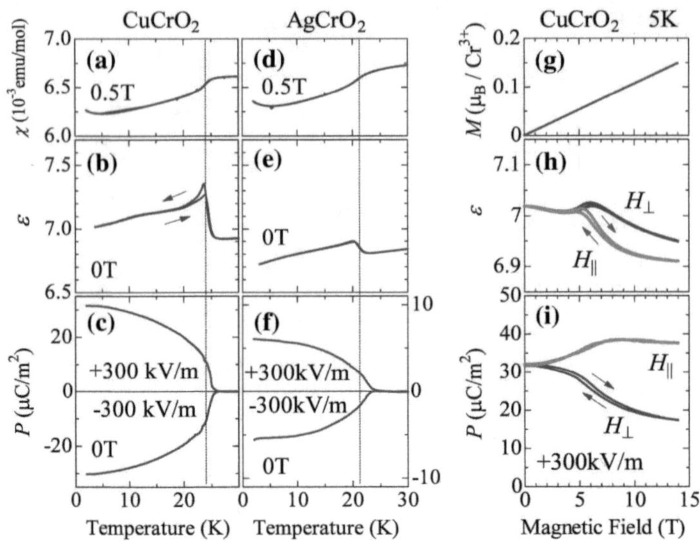

Fig. 3.27 Temperature dependence of magnetic susceptibility χ, dielectric constant ε, and electric polarization P for **a–c** $CuCrO_2$ and **d–f** $AgCrO_2$ with delafossite structure. **g–i** Magnetic field dependence of magnetization M, dielectric constant, and electric polarization for $CuCrO_2$. For **i**, electric poling was performed at 0 T. In **c**, **f** and **i**, the magnitude and sign of poling electric field are also indicated. $H_{\parallel}$ and $H_{\perp}$ indicate the magnetic field parallel and perpendicular to the direction of electric field

Results on Polycrystal

$ACrO_2$ (A = Cu, Ag, Li, or Na) as investigated here is viewed as another prototype of triangular lattice antiferromagnets. $CuCrO_2$ and $AgCrO_2$ crystallize into the delafossite structure (Fig. 3.26a), while $LiCrO_2$ and $NaCrO_2$ into the ordered rock salt structure (Fig. 3.26b). The both belong to the space group $R\bar{3}m$, and only a difference is in the stacking pattern of O-A-O layers; the delafossite structure has the straight stacking, while the ordered rock salt structure does the zigzag one. In the both cases the rhombohedral (ABCABC...) stacking is realized among Cr layers, although the distance between them is much shorter in the latter case [50]. The magnetic properties are dominated by Cr^{3+} ion with $S = 3/2$ spin. Because of the geometrical frustration of the antiferromagnetic interaction, the 120° spin structure is realized at the ground state. Former neutron diffraction studies have shown that these systems have the easy-axis anisotropy along the c-axis [51, 52].

Figure 3.27a–c shows the temperature dependence of magnetic susceptibility, dielectric constant, and electric polarization for $CuCrO_2$. The susceptibility shows a clear kink at $T_N \sim 24$ K, which signals the development of easy-axis type 120°-spin order with spiral plane including the c-axis [51]. At T_N, the dielectric constant also shows a sharp anomaly, and the spontaneous electric polarization begins to develop. With the opposite poling electric field, the polarization direction can be reversed.

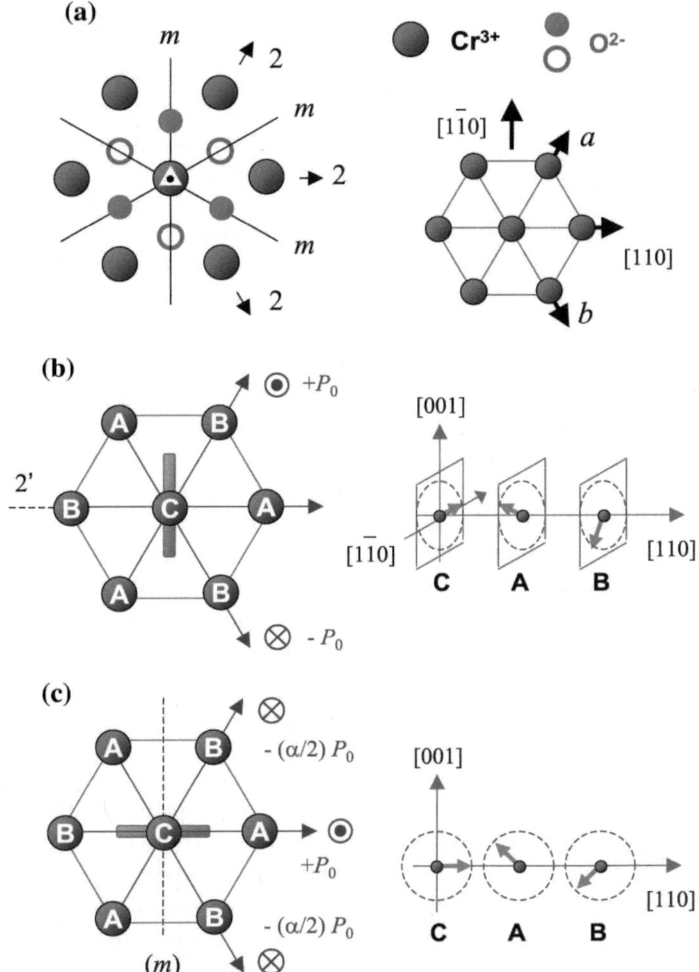

Fig. 3.28 **a** Symmetry elements at the magnetic Cr site (with $\bar{3}m$ site symmetry) in the stacked O-Cr-O triangular lattices: twofold rotation axis (2), reflection mirror (m), and threefold rotation axis along the c-axis with inversion center (*triangle* with *small circle*). O^{2-} site *above* (*below*) the Cr^{3+} layer is indicated as *closed* (*open*) *circle*. **b–c** Symmetry elements compatible with the 120° spin structure with **b** (110) spiral plane or **c** (1$\bar{1}$0) spiral plane. The *thick bars* (*orange online*, *left panel*) indicate the spin–spiral plane. Electric polarization expected from the spin-current model along each chain is also indicated, such as $\pm P_0$ and $-(\alpha/2)P_0$ (see text)

These indicate the ferroelectric nature of the ground state, coupled with the spiral magnetic order.

Figure 3.28a indicates the symmetry elements at magnetic Cr^{3+} site with site symmetry $\bar{3}m$. Hereafter, we examine two types of 120° magnetic order with spin–spiral either in the (110) plane (Fig. 3.28b) or in the (1$\bar{1}$0) plane (Fig. 3.28c). The former case can be considered as the proper screw magnetic structure, whose spins rotate in the plane perpendicular to the modulation vector. As discussed in "Introduction", proper screw magnetic order breaks several symmetry elements including the inversion center, and only the $2'$-axis perpendicular to the spin–spiral plane remains unbroken in the present lattice geometry. Thus, appearance of P is allowed only along the direction perpendicular to the spin–spiral plane. As for the microscopic origin of **P**, any 120° spin structure gives the same $\mathbf{S}_i \cdot \mathbf{S}_j$ for all bonds in the regular triangular lattice, and hence the conventional magnetostriction cannot cause the net polarization with centrosymmetric crystal structure. Another candidate, the inverse D–M model also predicts $P = 0$ for proper screw spin order. In case of $CuFe_{1-x}Ga_xO_2$ with delafossite crystal lattice and proper screw magnetic order, the spin-dependent modulation of covalency (hybridization) between metal d-state and ligand p-state, which also stems from the spin–orbit interaction, is suggested as the origin of ferroelectricity [9, 10]. Since $CuCrO_2$ possesses analogous crystallographic and magnetic geometry, it probably shares the same magnetoelectric coupling mechanism with $CuFe_{1-x}Ga_xO_2$.

In the case of $(1\bar{1}0)$ spiral plane (Fig. 3.28c), on the other hand, only a reflection mirror perpendicular both to the spin–spiral plane and to the triangular lattice basal plane can survive, depending on the direction of magnetic moment. Therefore, from the symmetry, polarization is allowed to appear along the direction perpendicular to the spin–spiral plane, or along the c-axis. The inverse D–M model predicts the polarization $(1-\alpha)P_0$ along the c-axis, where α represents the difference of coupling constant A_0 in Eq. (1) between the spin chains along [110] and [100] (or [010]). Given the isotropic coupling constant ($\alpha = 1$), the polarization should vanish, and hence other microscopic origin would be required. The similar argument as above is applicable for other centrosymmetric trigonal systems.

In Fig. 3.27g–i, magnetic field (H) dependence of magnetization, dielectric constant, and electric polarization for $CuCrO_2$ are indicated. The magnetization increases linearly with H up to 14 T. For dielectric properties, we performed measurements both under magnetic field parallel ($H_{\parallel}$) or perpendicular ($H_{\perp}$) to the direction of electric field. After the poling process at 0 T, electric polarization is enhanced with $H_{\parallel}$ and suppressed with $H_{\perp}$. The **P** value recovers when magnetic field is removed. In dielectric constant, corresponding anisotropic behavior between $H_{\parallel}$ and $H_{\perp}$ is observed above 6 T. These large magnetoelectric effects ensure the spin-driven ferroelectricity in this system. The threefold symmetry allows the existence of three equivalent magnetic domains. The large H-dependence of dielectric properties may be caused by the rotation and/or the volume change of the domains.

The isostructural material $AgCrO_2$ also shows the similar ferroelectricity driven by the magnetic order. Figure 3.27d–f indicate the temperature profiles of the same physical quantities for $AgCrO_2$. The kink in magnetic susceptibility is observed at $T_N \sim$21 K. Again, the anomaly in dielectric constant and the emergence of the ferroelectric polarization **P** are observed at T_N, although the **P** value is reduced as compared with $CuCrO_2$. A former powder neutron diffraction study has proposed a slightly modulated 120°-spin structure for the magnetic ground state below T_N, and the shorter correlation length and the larger spin fluctuation than in $CuCrO_2$ have been reported [53]. Although the detail of magnetic structure, such as the direction of

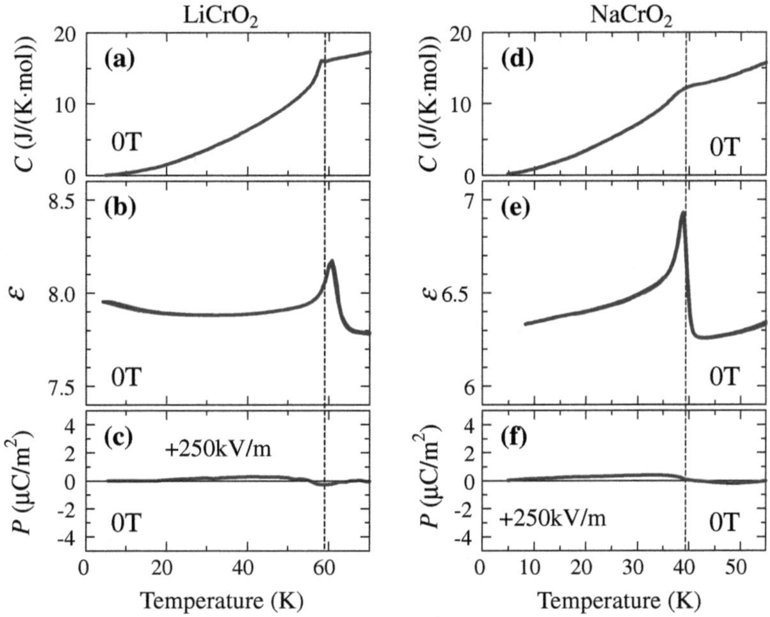

Fig. 3.29 Temperature profiles of specific heat capacity C, dielectric constant ε, and electric polarization P for **a–c** $LiCrO_2$ and **d–f** $NaCrO_2$ with ordered rock salt structure

spin–spiral plane, has not been determined yet, the smaller spontaneous polarization value in $AgCrO_2$ ($\sim 1/5$ of that for $CuCrO_2$) is consistent with these features. Since dielectric constant ε reflects the fluctuation of polarization (ΔP) in the form of $\varepsilon - \varepsilon_\infty \propto \langle |\Delta P|^2 \rangle / k_B T$, the weaker anomaly in ε must come from the smaller polarization.

In addition to the above delafossite crystals, we have also investigated $LiCrO_2$ and $NaCrO_2$ composed of the similar CrO_2 sheets but with ordered rock salt structure (Fig. 3.26b). The magnetic structure of $LiCrO_2$ has been investigated by the polarized neutron diffraction study on the single crystal [52], and below $T_N \sim 60$ K [54] the proper screw type 120°-spin structure (Fig. 3.28b) was reported to give the best fit. For $NaCrO_2$, only a powder neutron diffraction study was performed [55] and $T_N \sim 40$ K has been reported [56]. Figure 3.29a–f indicates the temperature profiles of heat capacity, dielectric constant, and electric polarization for $LiCrO_2$ and $NaCrO_2$. Although the anomaly in magnetic susceptibility is not clear [56], the heat capacity manifests the magnetic phase transitions, as seen in Fig. 3.29a and d, in accord with the former neutron diffraction studies. At T_N, dielectric constant shows a strong cusp like anomaly as in the two compounds with delafossite structure. Since dielectric constant is denoted as the fluctuation of electric polarization, the dielectric peak at T_N means the generation of the local electric dipole moments around the magnetic order temperature. This confirms the local correlation between the electric dipole and the magnetic structure also in this system. However, unlike the case for delafossites of $CuCrO_2$ and $AgCrO_2$, the macroscopic polarization can hardly be observed for $LiCrO_2$ or $NaCrO_2$. One of the possible interpretations for the absence of **P** but the presence of sharp ε-peak is the antiferroelectric order of electric dipoles. When ferroelectricity is induced by the spiral spin structure, the symmetry requires coupling between spin helicity (vector spin chirality as represented $\mathbf{S}_i \times \mathbf{S}_j$) and the sign of polarization. In fact, this relationship is confirmed by the recent polarized neutron scattering studies on several multiferroics (such as $TbMnO_3$ [57], $CuFe_{1-x}Al_xO_2$ [8]), on the basis of the two magnetic modulation vectors $\mathbf{q}_1 = (1/3, 1/3, 0)$ and $\mathbf{q}_2 = (-2/3, 1/3, 1/2)$, alternate stacking of Cr^{3+} layer with the opposite vector spin chirality was suggested [52]. Considering the spin-polarization coupling in $CuCrO_2$ and $AgCrO_2$, such an antiferro-chiral order in $LiCrO_2$ is naturally expected to induce the antiferroelectric state. For $CuCrO_2$ and $AgCrO_2$, by contrast, the $\mathbf{q}_2$ peaks, which characterize the alternate stacking of opposite chirality layers, have not been observed in neutron diffraction profiles [51, 53] in accord with the emergence of ferroelectricity in these compounds. At this stage, the origin of interaction that favors such antiferro-chiral spin order is an open question, since the inter-layer magnetic interaction should always favor the ferroic coupling of the vector spin chirality. The antiferroic coupling may possibly be ascribed to the inter-layer electrostatic and/or spin-lattice interaction. For example, the intra-layer polarization is governed by the spin chirality, yet could cause the lattice distortion that is dependent on the stacking form of the CrO_2 planes. The different stacking pattern of O-A-O layers and the shorter distance between Cr^{3+} layers in the ordered rock salt structure, which is anticipated to cause stronger inter-plane interaction and higher T_N [50], may be related to the antiferroic order of spin chirality.

Results on Single Crystal

Hereafter, we focus on the case for $CuCrO_2$. To analyze the anisotropy of magnetoelectric response, we investigated the nature of single crystal specimen. Figure 3.30a and b indicate the temperature dependence of ϵ and P. We can see that finite P appears only along the in-plane direction, not along out-of-plane direction (i.e. the c-axis). According to the latest polarized neutron scattering study on the single crystal, $CuCrO_2$ possesses spin–spiral plane parallel to the (110) plane. Thus, its magnetic structure is considered as a kind of proper screw, and the symmetry allows appearance of P perpendicular to the spin–spiral plane. The observation of in-plane P is consistent with this prediction.

Note that the genuine 120°-spin order with $q = 1/3$ possesses three magnetic q-vectors simultaneously (Fig. 3.25b), and thus does not produce the q-domains[10] unlike the case of $CuFe_{1-x}Ga_xO_2$ with $q \sim 0.202$. Still, there exist the choice of

[10] Recent neutron scattering study on single crystal reported $q \sim 0.329$ for $CuCrO_2$ [60], which slightly deviates from the ideal $q = 1/3$ of 120°-spin order. While this may allow the appearance of q-domains, the most of energy gain under applied magnetic field would still come from the reorientation of spin–spiral plane. Thus, in the following argument, we ignore the slight deviation

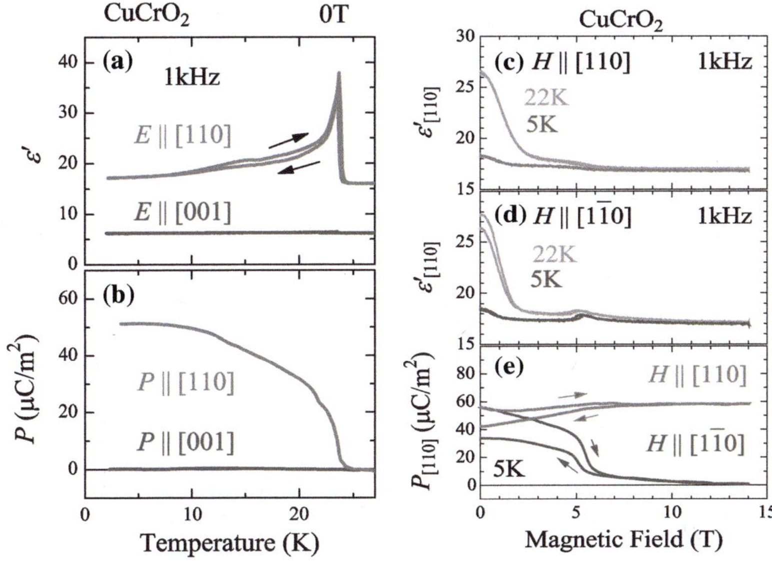

Fig. 3.30 Temperature dependence of **a** dielectric constant ϵ' and **b** electric polarization P for $CuCrO_2$ single crystal measured with $E \parallel [110]$ and $E \parallel [001]$ configurations, respectively. **c–e** Magnetic field dependence of $\epsilon_{[110]}$ and $P_{[110]}$ under $H_{[110]}$ or $H_{[1\bar{1}0]}$. For $P_{[110]}$, the specimen is first cooled at 0 T with poling E, and then measurement was performed without applied E in the process of H-scan (0 T → 14 T → 0 T). Note that recently similar results have also been reported by other groups [58, 59]

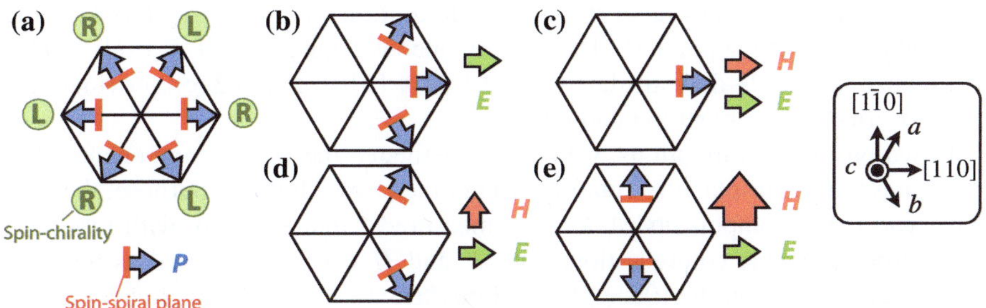

Fig. 3.31 **a** Six possible ferroelectric helimagnetic domains for $CuCrO_2$. *Blue arrows* and *red bars* indicate the P-direction and spin–spiral plane, respectively. *Circled* "R" and "L" denotes the chirality of spin–spiral. **b–e** Favorable domain configuration under applied external fields. With magnetic field larger than ∼5.5 T, the spin–spiral plane always becomes perpendicular to the applied H. In such large-H regions, the domain of spin–spiral plane vanishes and only the spin-chiral degree of freedom remains

three spin–spiral planes and two spin-chiral degree of freedom due to the symme-

---

from $q \sim 1/3$, which has recently been suggested to be related with the tiny local lattice distortion in the magnetic ordered phase [61]. The slight deviation from the $\theta_H \sim \theta_P$ relationship in Fig. 3.32b may be associated with this weak incommensurability and/or small lattice distortion.

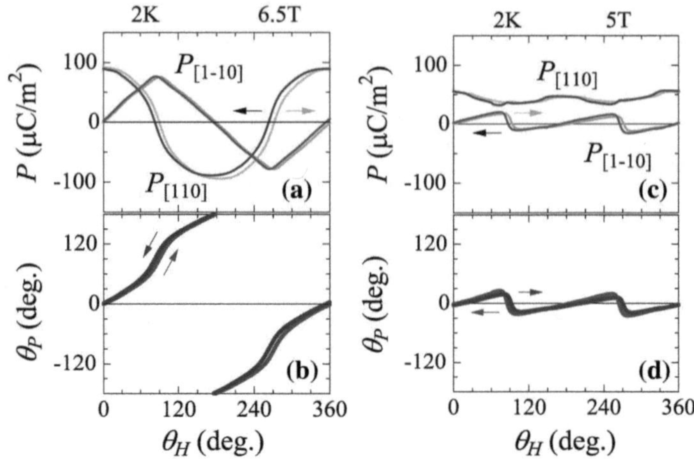

Fig. 3.32 [110] and [1$\bar{1}$0] components of P simultaneously measured under magnetic field rotating within the (001) plane with **a** $H = 6.5$ T and **c** $H = 5$ T, respectively. Here, we define θ_H (θ_P) as the angle between H- (P-) direction and the [110] axis (See Fig. 3.33). Arrows indicate the direction of H-rotation. Absolute value of P was determined by T-scan. The relationship between the H-direction and P-direction is plotted in (**b**) and (**d**)

try of triangular lattice, which leads to six possible P-domains with $P \parallel \langle 110 \rangle$ (Fig. 3.31a). As in the case for $CuFe_{1-x}Ga_xO_2$, electric field is expected to select the chirality of spin–spiral (Fig. 3.31b).[11] In contrast, magnetic field does not affect the spin-chirality but does favor the spin configuration where spin–spiral plane lies perpendicular to the applied H. Thus, simultaneous application of $E \parallel H \parallel [110]$ will realize the single-domain state with $P \parallel H \parallel [110]$ (Fig. 3.31c), provided that these external fields are strong enough.

Based on these assumptions, we have investigated the vector components of P in response to the in-plane H rotating around the c-axis. For this purpose, the both [110] and [1$\bar{1}$0] components of P were measured simultaneously with two pairs of electrodes. In this configuration, both P and H can be expressed as the two-dimensional vector on the (001) plane. Hereafter, we define the angle between H (P) and the [110]-axis as θ_H (θ_P). Since the specimen was cooled with H and E both applied along [110], we assume the uniform initial domain state as shown in Fig. 3.31c.

Figure 3.32a shows $P_{[110]}$ and $P_{[1\bar{1}0]}$ as a function of θ_H for the θ_H-increasing run, measured at $H = 6.5$ T without E. To see the development of P more directly, we plot the obtained θ_P against θ_H (Fig. 3.32b). In agreement with the expected initial state in Fig. 3.31c, $P \parallel H \parallel [110]$ is confirmed at $\theta_H = 0$. We can see that the relationship $\theta_P = \theta_H$ is mostly satisfied for all θ_H, which indicates that the P-vector smoothly rotates around the c-axis keeping $P \parallel H$. When we assume that

[11] The coupling between the spin-chirality and the sign of P, as well as E-control of spin-chirality, has recently been experimentally confirmed for $CuCrO_2$ by polarized neutron scattering study [62].

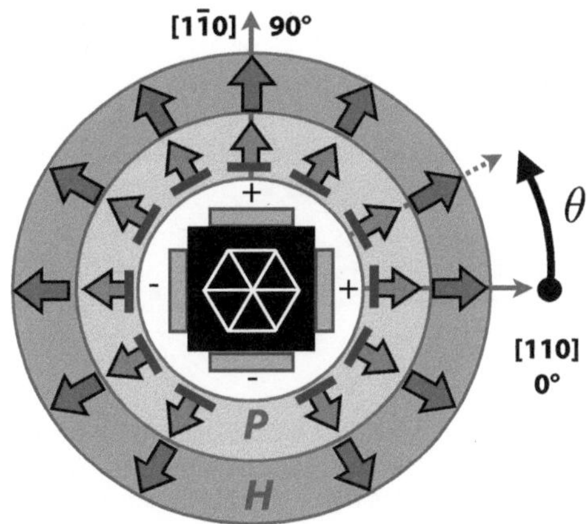

Fig. 3.33 (a) Relationship among P, H and spin–spiral plane for $CuCrO_2$. Here, *blue arrows* and *red bars* denote the P-direction and spin–spiral plane, respectively

the spin–spiral plane lies perpendicular to the applied magnetic field (i.e. spin–spiral plane rotates smoothly under rotating H), P turns out to always appear perpendicular to the spin–spiral plane (Fig. 3.33). Note that with either (110) or (1$\bar{1}$0) spin–spiral plane, the symmetry allows the appearance of in-plane P-component only along the direction perpendicular to the spin–spiral plane. The observed P-behavior well agrees with this prediction. The smooth rotation of P (and spin–spiral plane) for $CuCrO_2$ is in contrast with the case for $CuFe_{1-x}Ga_xO_2$, where discontinuous switching of P-vector (and q-vector) is observed for every 60° rotation of H.[12]

When we perform the same measurement at $H = 5$ T (Fig. 3.32c and d), θ_P shows periodic change with the cycle of 180°. Here, θ_P-value oscillates within the range of $30° > \theta_H > -30°$, suggesting the spin–spiral plane cannot slant from the original (110)-plane by more than ±30°. Since smooth rotation of spin–spiral plane requires the sufficiently large Zeeman energy to overcome the potential barrier arising from the magnetic anisotropy, the loss of the rotational behaviors of spin–spiral plane and associated P-vector would be reasonable in the lower-H region.

In Fig. 3.30c–e, the $H_{[110]}$- or $H_{[1\bar{1}0]}$-dependence of $\epsilon_{[110]}$ and $P_{[110]}$ are indicated. With increasing the magnitude of H, both $\epsilon_{[110]}$ and $P_{[110]}$ show clear anomaly around 5.5 T. Interestingly, this corresponds to the H-value which distinguishes the regions where spin–spiral plane can smoothly rotate or cannot. Thus, the anomaly at 5.5 T can be considered as the transition from the initial multiple-domain state at 0 T (Fig. 3.31b) into the single-domain state with spin–spiral plane perpendicular to H. Since $P \parallel H$ relationship is expected in higher-H region, observed suppression (enhancement) of $P_{[110]}$ under $H_{[1\bar{1}0]}$ ($H_{[110]}$) agrees well this scenario.

[12] In $CuFe_{1-x}Ga_xO_2$, such smooth rotation of spin–spiral plane cannot be observed. This implies that $CuFe_{1-x}Ga_xO_2$ possesses larger in-plane magnetic anisotropy than $CuCrO_2$.

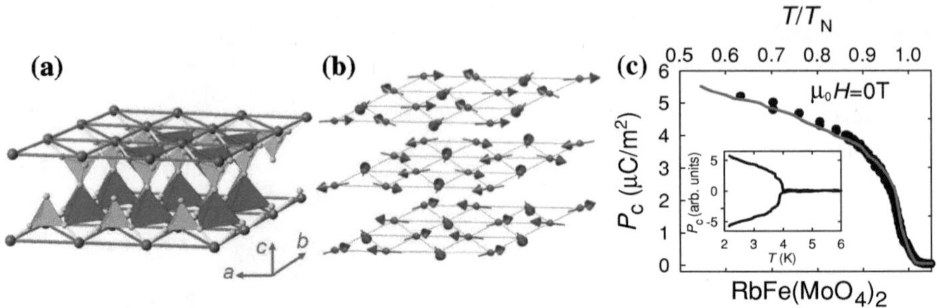

Fig. 3.34 **a** Crystal structure of triangular lattice antiferromagnet $RbFe(MoO_4)_2$ with space group $P\bar{3}$. **b** Magnetic ground state of $RbFe(MoO_4)_2$. **c** T-dependence of electric polarization along the c-axis. T_N indicates the Néel temperature. (Adapted with permission from [49], ©2007 APS.)

Discussion

As demonstrated above, the easy-axis type 120°-spin order on the trigonal lattice can induce the ferroelectric polarization along the direction perpendicular to the spin–spiral plane.

Interestingly, recent studies have also revealed that the easy-plane type 120°-spin order, where the spin–spiral lies within a plane parallel to the triangular lattice basal plane (i.e. *ab*-plane), can also induce ferroelectricity. Kenzelmann et al. investigated the dielectric properties and magnetic structure of triangular lattice antiferromagnet $RbFe(MoO_4)_2$, and reported the emergence of P along the c-axis upon the onset of easy-plane type 120°-spin order (Fig. 3.34) [49].

In Fig. 3.35a, we illustrate the symmetry elements compatible with the easy-plane type 120°-spin order. This magnetic order always possesses a threefold rotation axis (3) along the c-axis and m'(time-reversal followed by the mirror operation)-plane perpendicular to the c-axis. Depending on its in-plane spin direction, additional m'-plane perpendicular to $\langle 110 \rangle$ and twofold rotation axis (2) parallel to $\langle 1\bar{1}0 \rangle$ can appear. While this magnetic structure does not have the inversion center, the orthogonal arrangement of threefold rotation axis and m'-plane always prevents the system from being polar; Thus, the easy-plane type 120°-spin order cannot induce ferroelectricity in general. If the underlying crystal lattice does not have the mirror-plane parallel to the triangular lattice basal plane, however, the system can become polar along the c-axis and appearance of $P \parallel c$ is allowed. Notably, any trigonal (not hexagonal) lattice does satisfy this necessary condition.

In case of $RbFe(MoO_4)_2$ with space group $P\bar{3}$, the underlying crystal lattice possesses only the inversion center and threefold rotation axis along the c-axis. Thus, the inversion symmetry breaking by the easy-plane type 120°-spin order can create the polar axis along the c-axis [49]. As another example, we put this spin order on the stacked O-M-O triangular lattices and deduce the symmetry elements that remain unbroken in Fig. 3.35b. Even in this case, only $3 \parallel c$ axis and possible $m' \perp \langle 110 \rangle$ planes can survive, and hence emergence of $P \parallel c$ is allowed.

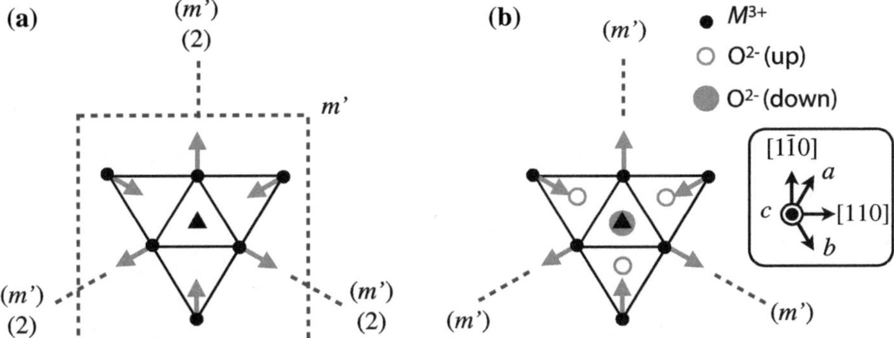

Fig. 3.35 **a** The symmetry elements compatible with the easy-plane type 120°-spin order; a threefold rotation axis (*black triangle*) along the c-axis and m'(time-reversal followed by the mirror operation)-plane perpendicular to the c-axis. Depending on its in-plane spin direction, additional m'-plane perpendicular to $\langle 110 \rangle$ and twofold rotation axis (2) parallel to $\langle 1\bar{1}0 \rangle$ can appear. **b** The easy-plane type 120°-spin order placed on the stacked O-M-O triangular lattices. Here, *open* (*closed*) *red circle* indicates the oxygen site *above* (*below*) the layer of magnetic M ion (*closed black circle*). The symmetry elements that remain unbroken are also indicated

Based on the above symmetry analysis, we can now predict that 120°-spin order on stacked triangular lattice can generally induce ferroelectricity, irrespective of their magnetic anisotropy. Here, P can always appear perpendicular to the spin–spiral plane. The present spin-driven ferroelectricity cannot be explained by either the exchange striction model or the inverse D–M model, which suggests that some other magnetoelectric coupling mechanism becomes active on the triangular lattice. Recently, Kan et al. performed the *ab initio* calculation based on the density functional theory for $AgCrO_2$ assuming the easy-axis type 120°-spin order with the (110) spin–spiral plane, and concluded that the spin–orbit interaction is indispensable to reproduce the emergence of ferroelectricity [63]. To fully understand the microscopic origin of magnetoelectric coupling on the triangular lattice, further theoretical efforts are highly desired.

Conclusion

In this section, we investigated the magnetoelectric response of triangular lattice antiferromagnet $ACrO_2$, and demonstrated that the easy-axis type 120°-spin order can induce ferroelectricity along the direction perpendicular to the spin–spiral plane. The observed ferroelectricity cannot be explained by either the exchange striction model or the inverse D–M model, which suggests that some other magnetoelectric coupling mechanism originating from spin–orbit interaction becomes active on the triangular lattice. Combined the present results with the recent report for $RbFe(MoO_4)_2$ with the easy-plane anisotropy [49], we can predict that a broad range

H																	He
2.20																	
Li	Be											B	C	N	O	F	Ne
0.98	1.57											2.04	2.55	3.04	3.44	3.98	
Na	Mg											Al	Si	P	S	Cl	Ar
0.93	1.31											1.61	1.90	2.19	2.58	3.16	
K	Ca	Sc	Ti	V	Cr	Mn	Fe	Co	Ni	Cu	Zn	Ga	Ge	As	Se	Br	Kr
0.82	1.00	1.36	1.54	1.63	1.66	1.55	1.83	1.88	1.91	1.90	1.65	1.81	2.01	2.18	2.55	2.96	3.00
Rb	Sr	Y	Zr	Nb	Mo	Tc	Ru	Rh	Pd	Ag	Cd	In	Sn	Sb	Te	I	Xe
0.82	0.95	1.22	1.33	1.6	2.16	1.9	2.2	2.28	2.20	1.93	1.69	1.78	1.96	2.05	2.1	2.66	2.60
Cs	Ba	*	Hf	Ta	W	Re	Os	Ir	Pt	Au	Hg	Tl	Pb	Bi	Po	At	Rn
0.79	0.89		1.3	1.5	2.36	1.9	2.2	2.20	2.28	2.54	2.00	1.62	2.33	2.02	2.0	2.2	2.2
Fr	Ra	**	Rf	Db	Sg	Bh	Hs	Mt	Ds	Rg	Cn	Uut	Uuq	Uup	Uuh	Uus	Uuo
0.7	0.9																
*	La	Ce	Pr	Nd	Pm	Sm	Eu	Gd	Tb	Dy	Ho	Er	Tm	Yb	Lu		
	1.1	1.12	1.13	1.14	1.13	1.17	1.2	1.2	1.1	1.22	1.23	1.24	1.25	1.1	1.27		
**	Ac	Th	Pa	U	Np	Pu	Am	Cm	Bk	Cf	Es	Fm	Md	No	Lr		
	1.1	1.3	1.5	1.38	1.36	1.28	1.13	1.28	1.3	1.3	1.3	1.3	1.3	1.3	1.3		

Fig. 3.36 Electronegativity deduced for each element (Pauling scale)

of trigonal materials with the 120°-spin structure can be multiferroic, irrespective of their magnetic anisotropy.

MX_2-Type Halides With CdI_2 Structure

Until now, the study of ferroelectric helimagnets is almost limited to oxides, partly because their isostructural chalcogen relatives (i.e. sulfides or selenides) are often too leaky to perform dielectric measurements [64–66]. Since halogens have larger electronegativity than chalcogens (Fig. 3.36), halides are expected to host better insulating nature and enable the investigation of ME properties for a wider variety of anions.

Among a variety of halide compounds, MX_2 (M = 3d transition metal, X = halogen) is known as one of the oldest and simplest examples in the history of magnetism. Depending on the choice of M and X, MX_2 takes three possible types of crystal structures (Fig. 3.37) [67]; TiO_2(rutile)-type structure is generally favored for X = F, and CdI_2- or $CdCl_2$-type structures are for X =Cl, Br, and I. Here, the latter two structures consist of the stacking of triangular lattices along the c-axis. They share the common MX_2 structural unit (as highlighted by blue squares in Fig.

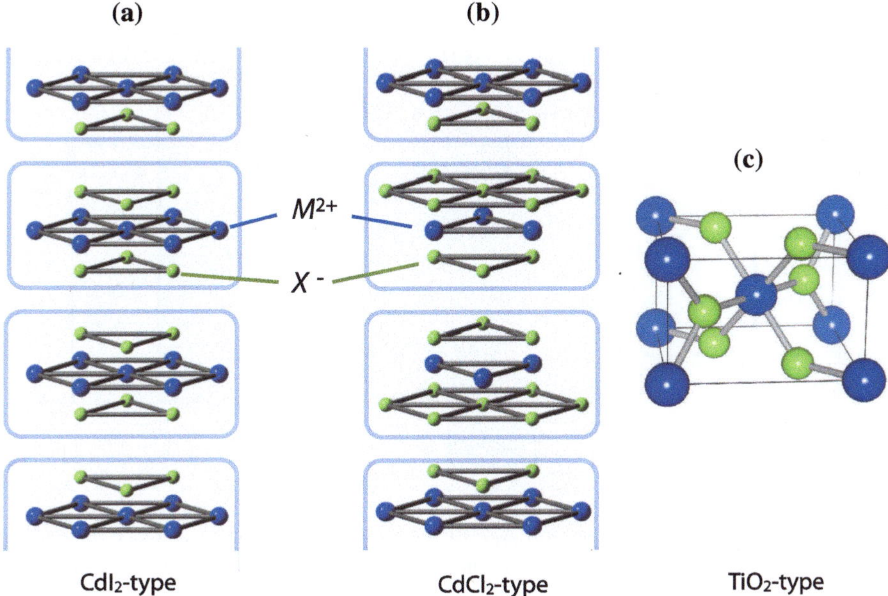

Fig. 3.37 Three typical crystal structures for MX_2-type halides

3.37a and b), and only the difference is its stacking pattern; CdI_2-type has the straight stacking while $CdCl_2$-type does the rhombohedral stacking. Reflecting the magnetic frustration on their triangular lattices, they frequently show the various types of spiral spin orders as summarized in Fig. 3.38.

In this chapter, we pick up three compounds with CdI_2-type structure (MnI_2 with proper screw spin order, VCl_2 with 120° spin order, and CoI_2 with cycloidal spin order), and investigated their magnetoelectric response in detail. These halides turn out to be the first examples of non-chalcogen based spiral-spin induced multiferroics, which promises further discovery of unique magnetoelectric response in other MX_2-type or different forms of halide compounds.[13]

MnI_2 with Proper Screw Spin Order

Introduction

MnI_2 is characterized by the CdI_2 structure as shown in Fig. 3.37a, and its magnetism is dominated by the Mn^{2+} ion with $S = 5/2$. According to Sato et al., MnI_2 undergoes three successive magnetic phase transitions at $T_{N1} \sim 3.95$K, $T_{N2} \sim 3.8$K, and $T_{N3} \sim$

[13] Note that for M = Cu (d^9) or Cr (d^4), Jahn Teller effect causes the distortion of the original crystal structure. The results for $CuCl_2$ with distorted triangular lattice are summarized in "$CuCl_2$".

	Ti (d^2)	V (d^3)	Cr (d^4)	Mn (d^5)	Fe (d^6)	Co (d^7)	Ni (d^8)	Cu (d^9)
F		Proper Screw 7.0K	AF 53K	AF 67K	AF 78K	AF 37K	Weak Ferro 73K	Weak Ferro 69K
Cl	AF 85K	120° *ac*-spiral 36K (1/3,1/3,1/2)	AF 16K	Complex 2K	AF 24K	AF 24K	AF 52K	*bc*-cycloid (1, 0.23, 0.5) 24K
Br		120° *ac*-spiral 30K (1/3,1/3,1/2)		↑↑↓↓ Spin ‖ *b* (1/4, 0, 1/4) 2K	AF 14K	AF 19K	*ab*-Cycloid (*q*,*q*,3/2) *q*=0.027 52K, 22.8K	cycloid? 24K
I		120°, AF *ac*-spiral 16K, 14K (1/3,1/3,1/2)		ProperScrew (0.18, 0, 0.4) 4.0K, 3.8K, 3.5K	AF 9K	*ab*-Cycloid (1/8, 0, 1/2) 11K, 9.4K	ProperScrew (0.14, 0, 0.15) 76K, 59K(?)	

CdI_2-type	Distorted CdI_2-type	Rutile (TiO_2)-type	Distorted TiO_2-type	$CdCl_2$-type
$P\bar{3}m1$	$C2/m$	$P4_2/mnm$	$P2_1/c$	$R\bar{3}m$

Fig. 3.38 Reported crystallographic and magnetic structures for various types of MX_2-type halides. Corresponding magnetic modulation vectors and transition temperatures are also indicated [67]

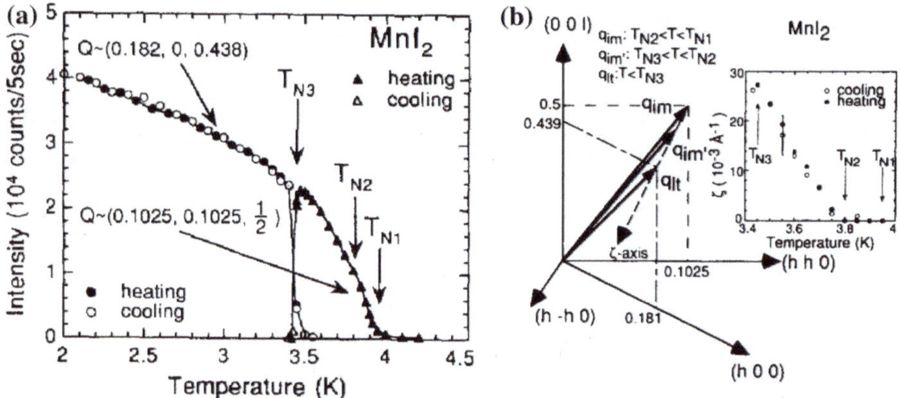

Fig. 3.39 Temperature variation of **a** intensity and **b** position of magnetic Bragg reflection observed for MnI_2. (Adapted with permission from [68], ©1995 Elsevier.)

3.45K (Fig. 3.39) [68]. Below T_{N1}, the magnetic Bragg reflection first appears at (0.1025, 0.1025, 1/2). With further decreasing temperature below T_{N2}, this reflection position begins to move slightly out of the (*hhl*)-plane towards the (*h*0*l*)-plane. Finally at T_{N3}, it jumps to $\mathbf{q} \sim (0.181, 0, 0.439)$.

So far, the magnetic structure is determined only for the magnetic ground state below T_{N3}. Previous neutron diffraction study concluded that the magnetic ground state of MnI_2 is the proper screw spin state, where spin rotates within a plane perpendicular to the modulation vector $\mathbf{q} \sim (0.181, 0, 0.439)$ [69]. Unlike the case for

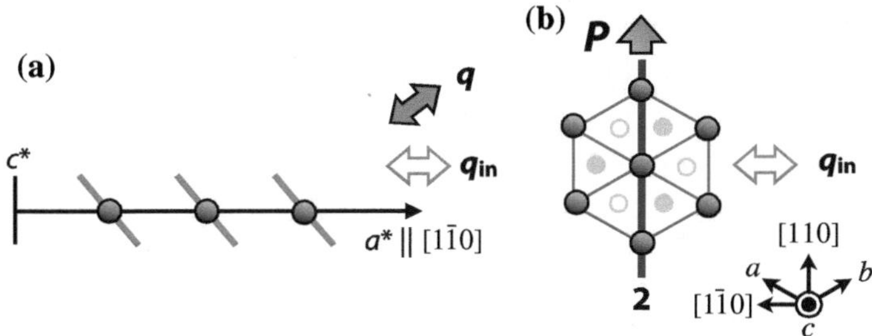

Fig. 3.40 Schematic illustration of proper screw magnetic ground state for MnI_2. q_{in} denotes the in-plane component of magnetic modulation vector q, and *red bars* in **a** indicate the spin–spiral plane perpendicular to the q-vector

proper screw spin state of $CuFe_{1-x}Ga_xO_2$ with $\mathbf{q} \sim (0.202, 0.202, 3/2)$ [18], the q-vector (i.e. spin–spiral axis) of MnI_2 is slanted from the triangular lattice basal plane (Fig. 3.40). In addition, the in-plane component of $\mathbf{q}$ ($\mathbf{q}_{in}$) for MnI_2 is pointing at the $[1\bar{1}0]$ direction, not the $[110]$ direction as observed for $CuFe_{1-x}Ga_xO_2$. Thus, the magnetic geometry for MnI_2 is quite different from the case for $CuFe_{1-x}Ga_xO_2$, and emergence of unique magnetoelectric response can be expected.

Result

Figure 3.41a–c indicates the temperature dependence of magnetic susceptibility χ, electric constant ϵ, and electric polarization P measured at 0 T for MnI_2. At $\sim$ 3.5 K, χ shows clear anomaly and only the in-plane component of ϵ shows a peak structure. Simultaneously, in-plane component of P begins to develop. Here, the P-direction was confirmed be reversed for the opposite direction of poling electric field. Since the present ferroelectric transition temperature agrees well with the reported $T_{N3} \sim 3.45$K, it is suggested that only the proper screw magnetic ground state (and not the intermediate magnetic phase between T_{N1} and T_{N3}) can induce the ferroelectricity of magnetic origin.

In Fig. 3.42a, the spin configuration in the magnetic ground state is schematically illustrated. While the original crystal lattice sustains a centrosymmetric $\bar{3}m$ site symmetry at magnetic Mn^{2+} site (Fig. 3.5b), proper screw spin order with $q_{in} \parallel [1\bar{1}0]$ breaks several symmetry elements including the inversion center. As a result, only the twofold rotation axis perpendicular both to the q-vector and the c-axis survives, and the emergence of $P \parallel [110] \perp q$ can be allowed (Fig. 3.42a) [9]. Interestingly, this is in contrast with the case for $CuFe_{1-x}Ga_xO_2$ with $q \parallel [110]$, where $P \parallel [110]$ appears "parallel" to the q-vector (Fig. 3.42b). The experimental observation of in-plane P (and no out-of-plane P) in MnI_2 agrees well with the above prediction. At this stage, it is not clear whether MnI_2 shares the same microscopic origin of magnetoelectric coupling with $CuFe_{1-x}Ga_xO_2$. Because of the canting of the spin–spiral plane towards the triangular lattice basal plane, the inverse D–M mechanism can also

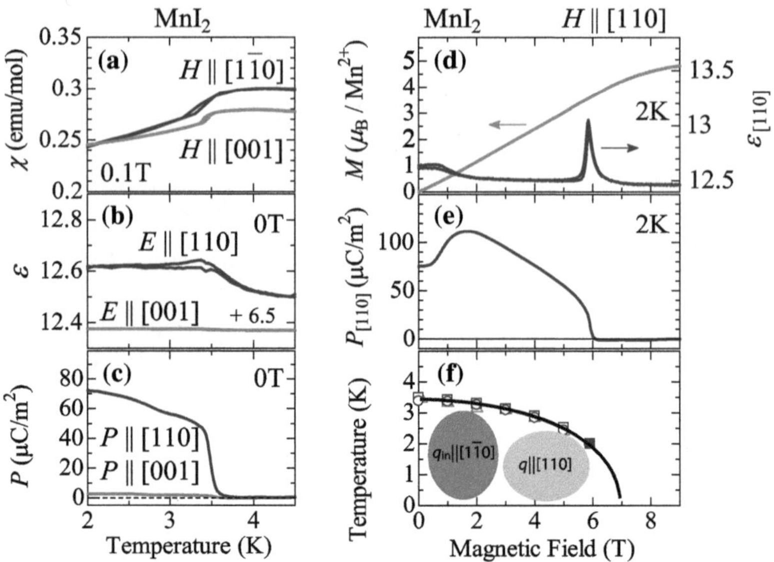

Fig. 3.41 Temperature dependence of **a** magnetic susceptibility χ, **b** dielectric constant ϵ, and **c** electric polarization P. **d** and **e** indicates $H_{[1\bar{1}0]}$-dependence of magnetization M, ϵ, and P. (f) $H-T$ phase diagram for $H \parallel [1\bar{1}0]$. *Triangles, circles, squares* are the data point obtained from the anomalies in M, ϵ, and P. *Open (closed) symbols* are take from T- (H-) increase runs

induce finite ferroelectric polarization along the direction predicted by the symmetry analysis.

Next, we investigate the behavior under magnetic field applied along the in-plane direction. Figure 3.41d and e indicates the H-dependence of magnetization M, [110] component of ϵ ($\epsilon_{[110]}$), and [110] component of P ($P_{[110]}$) measured under magnetic field along the $[1\bar{1}0]$ direction ($H_{[1\bar{1}0]}$). At 6 T, $\epsilon_{[110]}$ shows a sharp peak and $P_{[110]}$ vanishes completely. This probably corresponds to the magnetic transition from the helimagnetic phase to other magnetic phase, while no discernible anomaly can be observed in the $M-T$ curve. In Fig. 3.41f, the $H-T$ phase diagram determined from the various T- or H-scans of M, ϵ, and P is indicated. The boundary of ferroelectric phase always coincides with that of magnetic phase, which confirms the strong correlation between ferroelectricity and magnetism in this system.

As in case of $CuFe_{1-x}Ga_xO_2$, MnI_2 is also expected to host six equivalent ferroelectric domains with $P \parallel \langle 110 \rangle$, due to the symmetry of the underlying triangular lattice. In general, domains generated upon a magnetic transition must be converted to each other by the symmetry operation that is broken by the magnetic order.[14] By employing this rule, we can determine the corresponding q-direction and spin-chirality for each P-domain as summarized in Fig. 3.43a. Here, we can confirm that the $P \perp q_{in}$ relationship is always satisfied within each domain, and the reversal of

[14] The detail of this process is described in "Magnetic Digital Flop of Ferroelectric Domain".

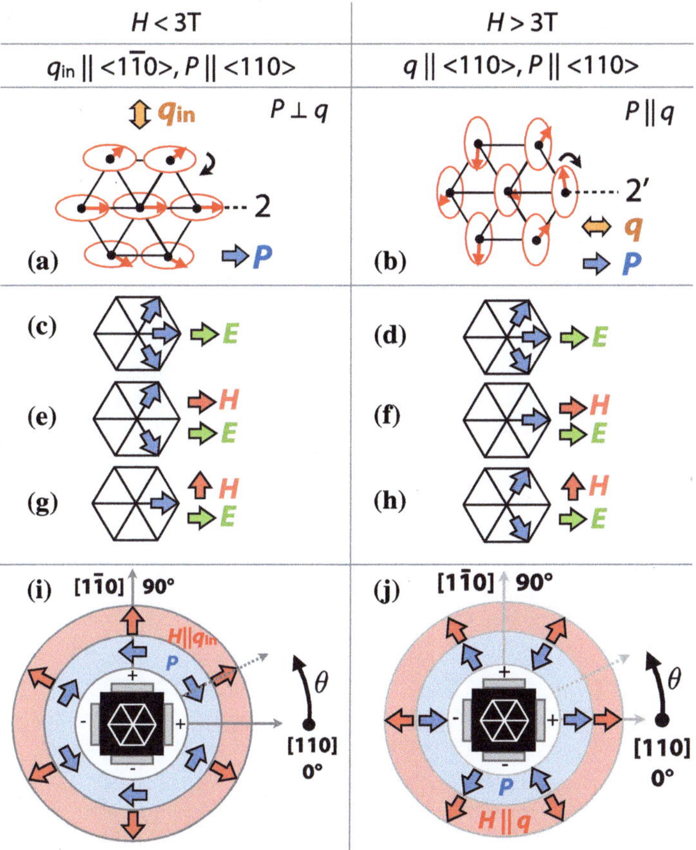

Fig. 3.42 **a** Proper screw magnetic ground state with $q_{\text{in}} \parallel \langle 1\bar{1}0 \rangle$, which is stable below 3 T. The compatible symmetry element and expected P-direction is also indicated. **b–d** Distribution of multiferroic domain(s) favored under various H and E. **e** Observed relationship among P, q, and H at $\theta_H = (30+60n)^\circ$ (n: integer). In the *right side* of the figure, corresponding charts for the proper screw magnetic phase with $q \parallel \langle 110 \rangle$, which appears above 3 T, are indicated. In **j**, the observed relationship among P, q, and H at $\theta_H = (60n)^\circ$ (n: integer) is illustrated

spin-chirality gives the opposite direction of P. From now, we consider the effect of external field on the domain distribution. Since the sign of P is governed by the spin-chirality, electric field is expected to select the chirality of spin–spiral (Fig. 3.42c).[15] In contrast, magnetic field does not lift the degeneracy of two spin-chiral domains, but is expected to favor the domain with antiferromagnetic moments most nearly perpendicular to the field direction. Consistently, the previous neutron scattering study on MnI_2 reported that the in-plane H ($\sim$1 T) selects the q-domain that possesses the relationship most close to $q \parallel H$ [69]. Thus, the simultaneous appli-

[15] Polarized neutron diffraction on $CuFe_{1-x}Ga_xO_2$ has confirmed the applied E affects only the chirality of spin–spiral, not the direction of q-vector.

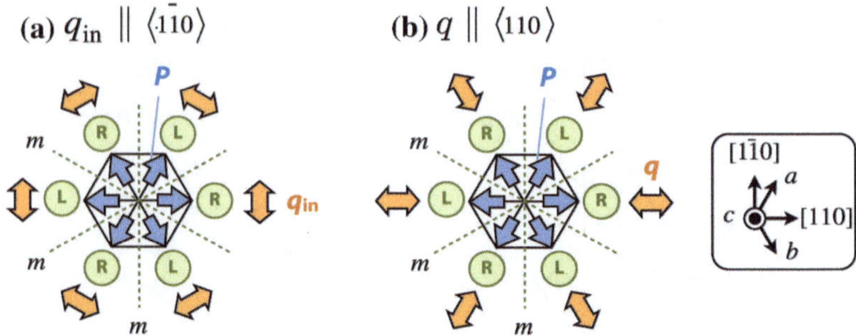

Fig. 3.43 Six possible P-domains and corresponding q-vector and spin chirality for **a** proper screw spin phase with $q_{in} \parallel \langle 1\bar{1}0 \rangle$ in the low-H region and **b** proper screw spin phase with $q \parallel \langle 110 \rangle$ in the high-H region, respectively. *Circled* "R" or "L" denote the chirality of spin–spiral. These domains can be converted into each other using the symmetry operation that is broken by the magnetic order. In the present case, for example, the application of mirror operation (m) to one specific domain gives another domain with reversed spin chirality

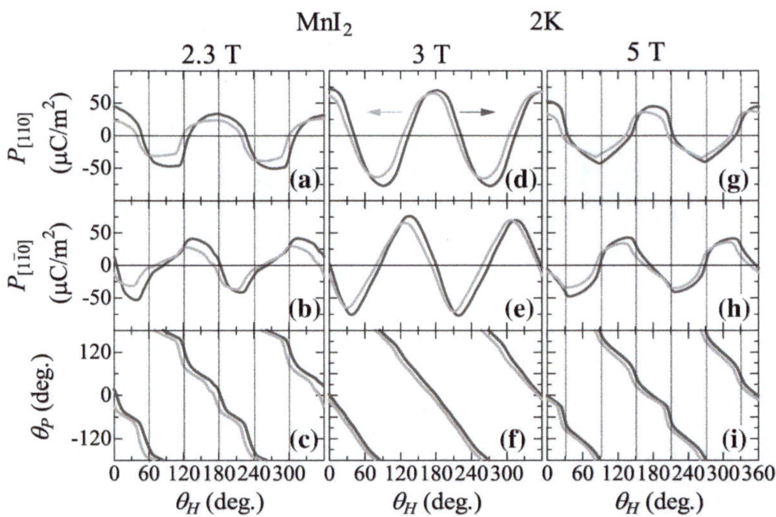

Fig. 3.44 **a**, **d**, **g** [110] and **b**, **e**, **h** [1$\bar{1}$0] components of P simultaneously measured in H rotating within the (001) plane. θ_H (θ_P) denotes the angle between H-vector (P-vector) and the [110] axis (see Fig. 3.42). In **c**, **f**, **i**, P-direction as a function of H-direction is also indicated. *Arrows* indicate the direction of H-rotation. Absolute value of P was determined by T-scan

cation of $E \parallel [110]$ and $H \parallel [1\bar{1}0]$ is expected to stabilize the single domain state with $P \parallel [110]$ (Fig. 3.42g).

Based on the above analysis, we have investigated the vector components of P in response to the in-plane H rotating around the c-axis. For this purpose, the both [110] and [1$\bar{1}$0] components of P were measured simultaneously with

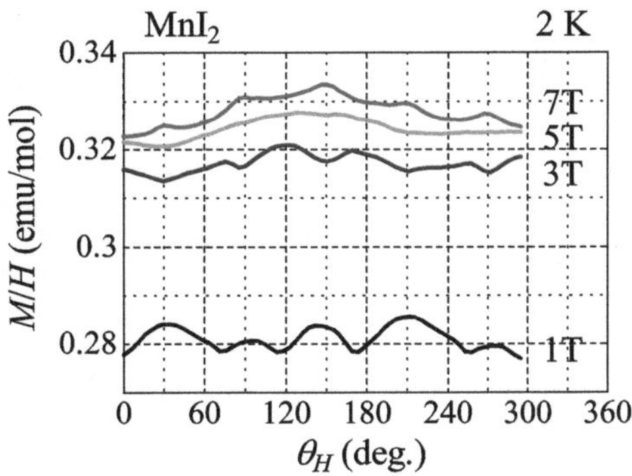

Fig. 3.45 H-direction dependence of magnetization. θ_H denotes the angle between H-vector and the [110] axis

two pairs of electrodes. In this configuration, both P and H can be expressed as the two-dimensional vector on the (001) plane. Hereafter, we define the angle between H (P) and the [110]-axis as θ_H (θ_P). Since the specimen was cooled with $H \parallel [1\bar{1}0]$ and $E \parallel -[110]$, we assume the uniform single domain state with $P \parallel -[110]$.

Figure 3.44a and b shows $P_{[110]}$ and $P_{[1\bar{1}0]}$ as a function of θ_H, measured at $H = 2.3$ T without E. Both $P_{[110]}$ and $P_{[1\bar{1}0]}$ show a periodic change with the cycle of 180°. To see the development of P more directly, we plot the obtained θ_P against θ_H (Fig. 3.44c). In agreement with the expected initial state, the relationship $H \perp P \parallel -[110]$ is confirmed at $\theta_H = 90°$. As θ_H increases, P suddenly flops by about 120° at $\theta_H \sim 120°$, and the relation $H \perp P \parallel [100]$ holds at $\theta_H = 150°$. Similar 120°-flop of P-vector is observed for every 60°-rotation of H, i.e. at $\theta_H = (60n)°$. In Fig. 3.42i, we summarized the observed P-direction at $\theta_H = (30 + 60n)°$. Here, we can see that the relationship $P \perp H$ always holds at each angle. Since H is expected to favor the domain with $H \parallel q_{in} \perp P$ and observed P-flop position ($\theta_H = (60n)°$) corresponds to the equilibrium point of two $q_{in} \parallel \langle 1\bar{1}0 \rangle$, the present H-induced P-flop transition must originate from the flop of q-vector.

We also performed the same measurement at $H = 5$ T.[16] Figure 3.44g–i indicate the θ_H-dependence of $P_{[110]}$, $P_{[1\bar{1}0]}$ and θ_P for $H = 5$ T. While 120°-flop of P-direction is again confirmed for every 60°-rotation of H, this P-flop behavior takes place at $\theta_H = (30 + 60n)°$, not at $\theta_H = (60n)°$ as in the case for $H = 2.3$ T. The observed P-direction at $\theta_H = (60n)°$ is summarized in Fig. 3.42j. At each θ_H, we can see that the relationship $P \parallel \pm H$ always hold. This is in contrast with the case for $H = 2.3$ T, where $P \perp H$ is favored. Interestingly, the present P-flop patterns under rotating H (~ 5 T) is perfectly the same as that observed for $CuFe_{1-x}Ga_xO_2$ in the proper screw spin state with $q \parallel \langle 110 \rangle$. This suggests that the proper screw magnetic ground state with $q_{\mathbf{in}} \parallel \langle 1\bar{1}0 \rangle$ (Fig. 3.42a) at 0 T is replaced by the proper

[16] For the P-profile at 5 T, the specimen was cooled with $H \parallel E \parallel [110]$ to obtain the uniform single domain state with $P \parallel [110]$ as shown in Fig. 3.42f.

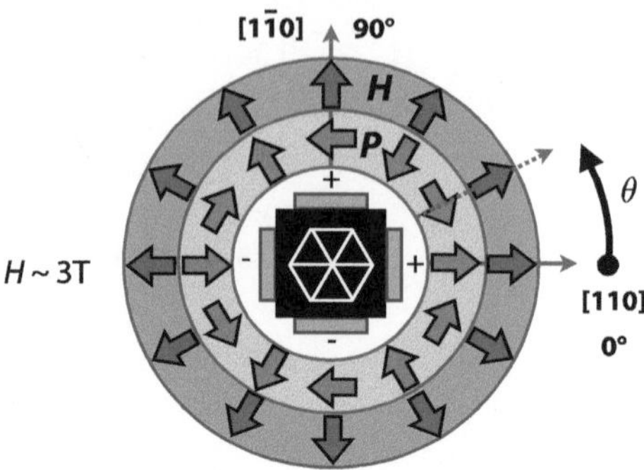

Fig. 3.46 Relationship between P and H at $\theta_H = (30n)°$ (n: integer) observed for $H = 3$ T

screw magnetic state with $q \parallel \langle 110 \rangle$ (Fig. 3.42b) in the higher-H region. To check this possibility, we measured M as a function of θ_H at various magnitude of magnetic field (Fig. 3.45). At 1 T, M takes minimum value at $\theta_H = (60n)°$, which agrees with the q-flop position assumed in the magnetic ground state. In contrast, M takes minimum value at $\theta_H = (30 + 60n)°$ at 3 T, which is rather consistent with the behavior expected for the proper screw spin order with $q \parallel \langle 110 \rangle$. The above results suggest that the change of stable q-direction from $q_{\mathrm{in}} \parallel \langle 1\bar{1}0 \rangle$ to $q \parallel \langle 110 \rangle$ probably takes place between $2.3 \sim 3$ T.

In Fig. 3.44d–f, θ_H-dependences of $P_{[110]}$, $P_{[1\bar{1}0]}$ and θ_P measured at $H = 3$ T are indicated.[17] Unlike the cases for 2.3 and 5 T, P-vector shows smooth rotation rather than discontinuous flop, keeping the relationship $\theta_P \sim -2\theta_H$. This means P-vector smoothly rotates clockwisely twice, when H-vector rotates counter-clockwisely only once. This is in contrast with the case for other ferroelectric helimagnets like $CuCrO_2$, where P, H and associated spin–spiral plane rotates toward the same direction with the same period. The observed P-directions at selected θ_H are summarized in Fig. 3.46. Interestingly, this P-profile at 3 T can be reproduced if we take the summation of the P-profiles observed at 2.3 T (Fig. 3.42i) and 5 T (Fig. 3.42i). Smooth rotation of P observed in the intermediate field region may reflect the continuous transition between $q_{\mathrm{in}} \parallel \langle 1\bar{1}0 \rangle$ and $q \parallel \langle 110 \rangle$.

We further investigated the variation of P-value under various strength of in-plane H. Figure 3.47a and c indicates the temperature dependence of $P_{[110]}$ under various magnitude of H applied parallel to the [1$\bar{1}$0] and [110] directions, respectively. Here, the specimen was first cooled with both H along selected direction and $E \parallel [110]$, and then the measurement was performed in the warming process keeping the applied H but without E. The obtained $P_{[110]}$-values at 2 K are plotted in Fig. 3.47b and d as open red circles. Since the domain with $H \parallel q_{\mathrm{in}} \perp P$ is favored in the low-H phase with $q_{\mathrm{in}} \parallel \langle 1\bar{1}0 \rangle$ and the one with $H \parallel q \parallel \pm P$ is favored in

[17] While the transition from $q_{\mathrm{in}} \parallel \langle 1\bar{1}0 \rangle$ phase to $q \parallel \langle 110 \rangle$ phase is completed below 3 T in M-profile, the corresponding transition field seems to be slightly higher in P-profile. This may come from the small deviation of temperature between these measurements.

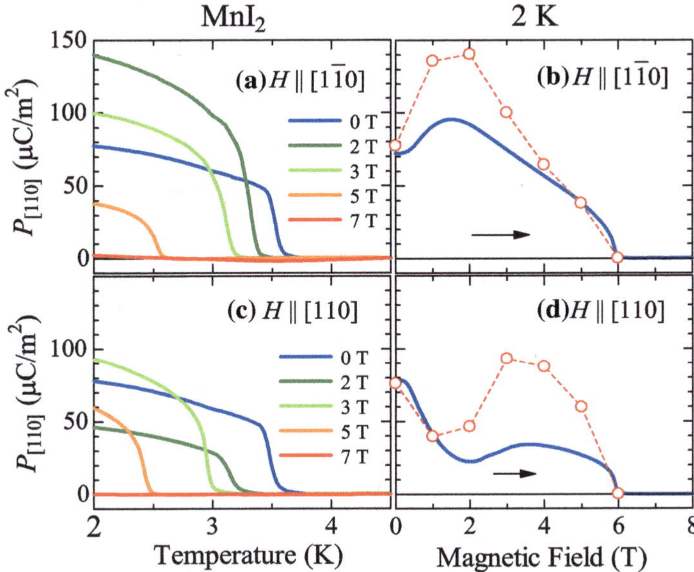

Fig. 3.47 Temperature dependence of $P_{[110]}$ under various magnitude of magnetic field applied along **a** $[1\bar{1}0]$ and **b** [110] direction, respectively. In **c** and **d**, the obtained P-values at 2 K are plotted with *red open circles* and *dashed line* as a function of H. For *blue solid line*, the specimen was first cooled at 0 T, and then measurement was performed in the H-increasing process. The deviation between *solid* and *dashed lines* probably reflects the loss of uniform spin-chirality during the process of magnetic phase transition and/or domain redistribution

the higher-H phase with $q \parallel \langle 110 \rangle$, we can expect the domain distribution as shown Fig. 3.42e–h for various configurations of E and H. With increasing the strength of $H \parallel [1\bar{1}0]$, the domain distribution goes through Fig. 3.42c → g → h. This predicts that $P_{[110]}$-value first increases, and then decreases above 3 T. The observed $H_{[1\bar{1}0]}$-dependence of $P_{[110]}$-value agrees well with this analysis (Fig. 3.47b). Likewise, with increasing the strength of $H \parallel [110]$, the domain distribution goes through Fig. 3.42c → e → f. This predicts that $P_{[110]}$-value first decreases, and then increases above 3 T. The observed $H_{[110]}$-dependence of $P_{[110]}$-value is also consistent with this analysis (Fig. 3.47d).

Finally, we discuss the behavior of spin-chirality upon the q-flop transition. When we compare the Figs. 3.42i and 3.43a, we can see that the spin-chirality is always preserved upon the q-flop in the $q_{in} \parallel \langle 1\bar{1}0 \rangle$ state. Likewise, the comparison between Figs. 3.42j and 3.43b also suggests the preservation of spin-cirality upon the q-flop in the $q \parallel \langle 110 \rangle$ state. Since the degeneracy of two spin-chiral states cannot be lifted under the applied H, the selection of odd chirality upon the q-flop must reflect the nature of multiferroic domain wall as already discussed in "Magnetic Digital Flop of Ferroelectric Domain" . Combined with the case for $CuFe_{1-x}Ga_xO_2$, such preservation of spin-chirality upon q-flop (or stability of 120° domain wall rather than 60° domain wall) may be the common nature on the triangular lattice with proper screw spin order. To fully interpret the presently observed magnetoelectric response, further theoretical studies on the nature of multiferroic domain wall are highly desired.

Conclusion

In this study, we investigated the magnetoelectric response of triangular lattice antiferromagnet MnI_2. We discovered the ferroelectric nature of the proper screw magnetic ground state, and in-plane H was found to induces the rearrangement of the six possible multiferroic domains in this compound. With every 60°-rotation of H around the c-axis, 120°-flop of P-vector is observed as a result of the flop of magnetic q-vector. The stable q-direction changes from the original $[1\bar{1}0]$ to [110] above 3 T, which leads to the 30°-shift of H-angle where q-flop transition takes plane. At the intermediate field region (~3 T), the clockwise rotation of H by 360° leads to smooth counter-clockwise rotation of P by 720°. As observed in the case of $CuFe_{1-x}Ga_xO_2$, the chirality of spin–spiral is always preserved upon the flop of q-vector, which implies the stability of specific form of multiferroic domain wall.

CoI_2 With Cycloidal Spin Order

CoI_2 is characterized by the CdI_2 structure, and its magnetism in is dominated by the Co^{2+} ion with $S = 3/2$. According to Kuindersma et al., this compound hosts cycloidal spin order with spin–spiral plane confined in the ab-plane and magnetic modulation vector $q \sim (1/8, 0, 1/2)$, below $T_N \sim 8$ K [70].

Figure 3.48 indicates the temperature dependence of χ, ϵ and P measured under 0 T for CoI_2. At $T_N \sim 8$ K, χ shows clear anomaly and in-plane component of P begins to develop. The inverse D–M model predicts the appearance of in-plane P along the direction perpendicular to the q-direction. The observation of in-plane P is consistent with this prediction.

VCl_2 With 120°-Spin Order

VCl_2 is characterized by the CdI_2 structure, and its magnetism is dominated by the V^{2+} ion with $S = 3/2$. This compound hosts the easy-axis type 120°-spin order with magnetic modulation vector $q = (1/3, 1/3, 1/2)$ below $T_N \sim 36$ K [71].

Figure 3.49 indicates the temperature dependence of ϵ and P measured with $E \parallel c$ for VCl_2. At $T_N \sim 36$ K, ϵ shows clear anomaly and P begins to develop. Since this compound possesses the same magnetic geometry with $CuCrO_2$, the emergence of in-plane P is expected. The observed P_c component may come from the leakage

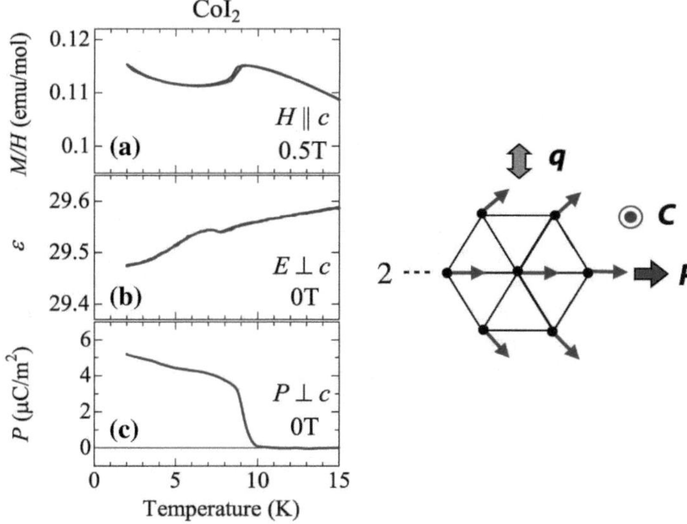

Fig. 3.48 Temperature dependence of χ, ϵ and P measured for CoI_2. Reported helimagnetic structure is also illustrated. Here, we define the vector $\mathbf{C} = \mathbf{S}_i \times \mathbf{S}_j$, which is perpendicular to the spin–spiral plane

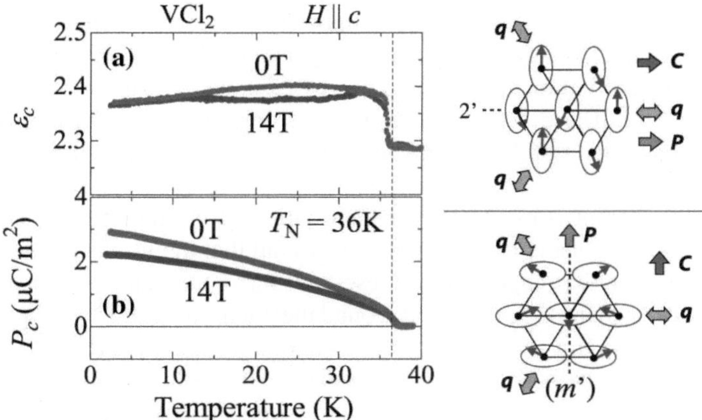

Fig. 3.49 Temperature dependence of χ, ϵ and P measured for VCl_2. Two possible easy-axis type 120°-spin orders are also illustrated. Here, we define the vector $\mathbf{C} = \mathbf{S}_i \times \mathbf{S}_j$, which is perpendicular to the spin–spiral plane

from the in-plane P. Because of the thin plate shape of the specimen, it is difficult to measure the anisotropy of magnetoelectric response at this stage. The growth of thicker specimen and further investigation is required.

Summary

In this chapter, we investigated the magnetoelectric response of the triangular lattice antiferromagnets. Our measurements prove that various types of proper screw spin order and 120°-spin order on trigonal lattice can indeed induce ferroelectricity, while both exchange striction and inverse D–M model denies the appearance of electric polarization for these spin textures. This suggests that some other magnetoelectric coupling mechanism originating from spin–orbit coupling is active on the stacked triangular lattices, which may be relevant with the our observation of novel electromagnon excitation in the collinear spin state. The symmetry of triangular lattice also hosts multiple multiferroic domains with different P-directions, which offers unique magnetoelectric responses under rotating magnetic field as demonstrated. In the following, we briefly summarize the results obtained in each section.

$CuFeO_2$ With Proper Screw Spin Order (q ∥ [110])

- First, we investigated the effect of nonmagnetic impurity (Al^{3+} or Ga^{3+}) doping on the magnetic Fe^{3+} site. Reflecting its competing magnetic ground states, even a small amount of nonmagnetic impurity drastically changes the magnetic phase diagram. Especially, the critical magnetic field necessary for the induction of ferroelectric helimagnetic phase decreases down to zero with Al-doping up to $x = 0.02$. These results prove that not only magnetic field but also site-dilution can often stabilize magnetic ferroelectricity via the modification of spin frustration.
- Next, we attempted to control the distribution of six multiferroic domains in $CuFe_{1-x}Ga_xO_2$ using external field. We found that the in-plane magnetic field can induce the rearrangement of these multiferroic domains. Upon every 60°-rotation of in-plane magnetic field around the c-axis, unique 120°-flop of electric polarization occurs as a result of the switch of helical magnetic q-vector. The chirality of spin helix is always conserved upon the q-flop, which implies that the nature of multiferroic domain wall plays an important role in the determination of P-behavior under magnetic field.
- Finally, we investigated the dynamics of this compound. Terahertz time-domain spectroscopy was performed to directly probe the low-energy (1–5 meV) electrodynamics of $CuFe_{1-x}Ga_xO_2$. We discovered an unique electromagnon (electric-field-active magnon) excitation at 2.3 meV in the paraelectric ↑↑↓↓ collinear magnetic phase, and this electromagnon vanishes in the ferroelectric helimagnetic phase. While previously discovered electromagnon excitations in other materials are believed to arise from the exchange striction mechanism, the present anti-correlation with noncollinear magnetism excludes the exchange-striction mechanism as the origin of dynamical magnetoelectric coupling in

$CuFe_{1-x}Ga_xO_2$. This evidences the first experimental observation of spin–orbit coupling mediated electromagnon in the present compound.

ACrO$_2$ With 120°-Spin Order

In this section, we investigated the magnetoelectric response of 120°-spin order, which was previously predicted to be paraelectric according to the inverse D–M model. Our targets are $ACrO_2$ (A = Cu, Ag) with the easy-axis type 120°-spin order, and we demonstrated that these compounds can induce ferroelectricity along the direction perpendicular to the spin–spiral plane. Combined the present results with the recent report for $RbFe(MoO_4)_2$ with the easy-plane anisotropy [49], we can predict that a broad range of trigonal materials with the 120°-spin structure can be multiferroic, irrespective of their magnetic anisotropy.

MnI$_2$ With Proper Screw Spin Order (q_{in} ∥ [110])

MnI_2 is characterized by the proper screw magnetic ground state, but with $q_{in} \parallel \langle 1\bar{1}0\rangle$ unlike the case of $CuFe_{1-x}Ga_xO_2$ with $q \parallel \langle 110\rangle$. We discovered the ferroelectric nature of this helimagnetic ground state, and in-plane H was found to induces the rearrangement of the six possible multiferroic domains. Here, the relationship $P \perp q_{in} \parallel H$ is always favored, which is in contrast with the case of $CuFe_{1-x}Ga_xO_2$ with $P \parallel q \parallel H$. With every 60°-rotation of H around the c-axis, 120°-flop of P-vector is observed as a result of the flop of magnetic q-vector. Interestingly, the stable q-direction changes from the original $[1\bar{1}0]$ to [110] above 3 T, which leads to the 30°-shift of H-angle where q-flop transition takes plane. As in the case of $CuFe_{1-x}Ga_xO_2$, the chirality of spin–spiral is always preserved upon the flop of q-vector, which again implies the stability of specific form of multiferroic domain wall.

Other MX_2-Type Halides (CoI$_2$ and VCl$_2$)

We further investigated the magnetoelectric response in other MX_2-type triangular lattice halides; CoI_2 with cycloidal spin order and VCl_2 with easy-axis type 120°-spin order. In both compounds, we discovered the emergence of magnetically-induced ferroelectricity. MX_2 is the first example of non-chalcogen based spiral-spin induced multiferroics, while so far the study of ferroelectric helimagnets is almost limited to oxides. This is partly because their isostructural chalcogen relatives with larger anions (i.e. sulfides or selenides) are often electrically too leaky to perform dielectric measurements. Since halogens have larger electronegativity than chalcogens, halides are better insulating and enable the investigation of ME properties for a wider variety of anions as clearly demonstrated here. Our present results promises further discovery of unique ME function in many MX_2-type compounds and other forms of halide compounds.

References

1. Collins MF, Petrenko OA (1997) Triangular antiferromagnets. Can J Phys 75:605
2. Kimura T, Lashley JC, Ramirez AP (2006) Inversion-symmetry breaking in the noncollinear magnetic phase of the triangular-lattice antiferromagnet $CuFeO_2$. Phys Rev B 73:220401(R)
3. Zhao TR, Hasegawa M, Takei H (1996) Crystal growth and characterization of cuprous ferrite ($CuFeO_2$). J Cryst Growth 166:408
4. Mitsuda S, Mase M, Prokes K, Kitazawa H, Katori HA (2000) Field-induced magnetic phase transitions in a triangular lattice antiferromagnet $CuFeO_2$ up to 14.5 T. J Phys Soc Jpn 69:3513
5. Mitsuda S, Yoshizawa H, Yaguchi N, Mekata M (1991) Neutron diffraction study of $CuFeO_2$. J Phys Soc Jpn 60:1885
6. Nakajima T, Mitsuda S, Kanetsuki S, Prokes K, Podlesnyak A, Kimura H, Noda Y (2007) Spin noncollinearlity in multiferroic phase of triangular lattice antiferromagnet $CuFe_{1-x}Al_xO_2$. J Phys Soc Jpn 76:043709
7. Mitsuda S, Kasahara N, Uno T, Mase M (1998) Partially disordered phase in frustrated triangular lattice antiferromagnet $CuFeO_2$. J Phys Soc Jpn 67:4026
8. Nakajima T, Mitsuda S, Kanetsuki S, Tanaka K, Fujii K, Terada N, Soda M, Matsuura M, Hirota K (2008) Electric polarization induced by a proper helical magnetic ordering in a delafossite multiferroic $CuFe_{1-x}Al_xO_2$. Phys Rev B 77:052401
9. Arima T (2007) Ferroelectricity induced by proper-screw type magnetic order. J Phys Soc Jpn 76:073702
10. Jia C, Onoda S, Nagaosa N, Han JH (2006) Bond electronic polarization induced by spin. Phys Rev B 74:224444
11. Terada N, Mitsuda S, Prokes K, Suzuki O, Kitazawa H, Katori HA (2004) Impact of a small number of nonmagnetic impurities on H-T magnetic phase diagram of $CuFeO_2$. Phys Rev B 70:174412
12. Terada N, Kawasaki T, Mitsuda S, Kimura H, Noda Y (2005) Reinvestigation of magnetic structures for the thermally induced states of $CuFe_{1-x}Al_xO_2$ (x =0.00, 0.02 and 0.05) using a four-circle neutron diffractometer. J Phys Soc Jpn 74:1561
13. Terada N, Mitsuda S, Suzuki S, Kawasaki T, Fukuda M, Nagao T, Katori HA (2004) Disappearance of quasi-ising character in triangular lattice antiferromagnet $CuFeO_2$ by a small amount of substitution. J Phys Soc Jpn 73:1442
14. Burgy J, Mayr M, Martin-Mayor V, Moreo A, Dagotto E (2001) Colossal effects in transition metal oxides caused by intrinsic inhomogeneities. Phys Rev Lett 87:277202
15. Terada N, Mitsuda S, Fujii T, Soejima K, Doi I, Katori HA, Noda Y (2005) Magnetic phase diagram of the triangular lattice antiferromagnet $CuFe_{1-x}Al_xO_2$. J Phys Soc Jpn 74:2604
16. Terada N, Nakajima T, Mitsuda S, Kitazawa H (2009) Magnetic phase diagram of multiferroic delafossite $CuFe_{1-y}Ga_yO_2$. J Phys Conf Ser 145:012071
17. Nakajima T, Mitsuda S, Takahashi K, Yamano M, Masuda K, Yamazaki H, Prokes K, Kiefer K, Gerischer S, Terada N, Kitazawa H, Matsuda M, Kakurai K, Kimura H, Noda Y, Soda M, Mitsuda M, Hirota K (2009) Comprehensive study on ferroelectricity induced by a proper-screw-type magnetic ordering in multiferroic $CuFeO_2$: nonmagnetic impurity effect on magnetic and ferroelectric order. Phys Rev B 79:214423
18. Terada N, Nakajima T, Mitsuda S, Kitazawa H, Kaneko K, Metoki N (2008) Ga-substitution-induced single ferroelectric phase in multiferroic $CuFeO_2$. Phys Rev B 78:014101
19. Terada N, Mitsuda S, Fujii T, Petitgrand D (2007) Inelastic neutron scattering study of frustrated Heisenberg triangular magnet $CuFeO_2$. J Phys Condens Matter 19:145241

20. Ye F, Fernandez-Baca JA, Fishman RS, Ren Y, Kang HJ, Qiu Y, Kimura T (2007) Magnetic interactions in the geometrically frustrated triangular lattice antiferromagnet $CuFeO_2$. Phys Rev Lett 99:157201
21. Fishman RS (2008) Spin waves in $CuFeO_2$. J Appl Phys 103:07B109
22. Takagi T, Mekata M (1995) New partially disordered phases with commensurate spin density wave in frustrated triangular lattice. J Phys Soc Jpn 64:4609
23. Haraldsen JT, Swanson M, Alvarez G, Fishman RS (2009) Spin-wave instabilities and noncollinear magnetic phases of a geometrically frustrated triangular-lattice antiferromagnet. Phys Rev Lett 102:237204
24. Adachi K (1996) Magnetism of compounds—localized spin system. Shokabo, Tokyo
25. Terada N, Nakajima T, Mitsuda S, Tanaka Y, Mamiya H, Kitazawa H (2010) Charge disproportionation associated with spin ordering in delafossite $CuFeO_2$ as seen via resonant x-ray diffraction. Phys Rev B 81:064424
26. Ascher E, Rieder H, Schmid H, Sössel H (1966) Some properties of ferromagnetoelectric nickel–iodine boracite, $Ni_3B_7O_{13}I$. J Appl Phys 37:1404
27. Hill NA (2000) Why are there so few magnetic ferroelectrics? J Phys Chem B 104:6694
28. Schmid H (2008) Some symmetry aspects of ferroics and single phase multiferroics. J Phys Condens Matter 20:434201
29. Gajek M, Bibes M, Fusil S, Bouzehouane K, Fontcuberta J, Barthelemy A, Fert A (2007) Tunnel junctions with multiferroic barriers. Nat Mater 6:296
30. Yamasaki Y, Miyasaka S, Kaneko Y, He J-P, Arima T, Tokura Y (2006) Magnetic reversal of the ferroelectric polarization in a multiferroic spinel oxide. Phys Rev Lett 96:207204
31. Abe N, Taniguchi K, Ohtani S, Takanobu T, Iwasa Y, Arima T (2007) Polarization reversal in multiferroic $TbMnO_3$ with a rotating magnetic field direction. Phys Rev Lett 99:227206
32. Kagawa F, Mochizuki M, Onose Y, Murakawa H, Kaneko Y, Furukawa N, Tokura Y (2009) Dynamics of multiferroic domain wall in spin-cycloidal ferroelectric $DyMnO_3$. Phys Rev Lett 102:057604
33. Taniguchi K, Abe N, Umetsu H, Katori HA, Arima T (2008) Control of the magnetoelectric domain-wall stability by a magnetic field in a multiferroic $MnWO_4$. Phys Rev Lett 101:207205
34. Abe N, Taniguchi K, Ohtani S, Umetsu H, Arima T (2009) Control of the polarization flop direction by a tilted magnetic field in multiferroic $TbMnO_3$. Phys Rev B 80:020402(R)
35. Murakawa H, Onose Y, Ohgushi K, Ishiwata S, Tokura Y (2008) Generation of electric polarization with rotating magnetic field in helimagnet $ZnCr_2Se_4$. J Phys Soc Jpn 77:043709
36. Katsura H, Nagaosa N, Balatsky AV (2005) Spin current and magnetoelectric effect in noncollinear magnets. Phys Rev Lett 95:057205
37. Pimenov A, Mukhin AA, Ivanov VYu, Travkin VD, Balbashov AM, Loidl A (2006) Possible evidence for electromagnons in multiferroic manganites. Nat Phys 2:97
38. Kida N, Ikebe Y, Takahashi Y, He JP, Kaneko Y, Yamasaki Y, Shimano R, Arima T, Nagaosa N, Tokura Y (2008) Electrically driven spin excitation in the ferroelectric magnet $DyMnO_3$. Phys Rev B 78:104414
39. Sushkov AB, Aguilar RV, Park S, Cheong S-W, Drew HD (2007) Electromagnons in Multiferroic YMn_2O_5 and $TbMn_2O_5$. Phys Rev Lett 98:027202
40. Kida N, Okuyama D, Ishiwata S, Taguchi Y, Shimano R, Iwasa K, Arima T, Tokura Y (2009) Electric-dipole-active magnetic resonance in the conical-spin magnet $Ba_2Mg_2Fe_{12}O_{22}$. Phys Rev B 80:220406(R)
41. Ishiwata S, Taguchi Y, Murakawa H, Onose Y, Tokura Y (2008) Low-magnetic-field control of electric polarization vector in a helimagnet. Science 319:1643
42. Katsura H, Balatsky AV, Nagaosa N (2007) Dynamical magnetoelectric coupling in helical magnets. Phys Rev Lett 98:027203
43. Sushkov AB, Mostovoy M, Aguilar RV, Cheong S-W, Drew HD (2008) Electromagnons in multiferroic RMn_2O_5 compounds and their microscopic origin. J Phys Condens Matter 20:434210

44. Aguilar RV, Mostovoy M, Sushkov AB, Zhang CL, Choi YJ, Cheong S-W, Drew HD (2009) Origin of electromagnon excitations in multiferroic RMnO$_3$. Phys Rev Lett 102:047203
45. Miyahara S, Furukawa N (2008) Theory of electric field induced one-magnon resonance in cycloidal spin magnets. arXiv:0811.4082
46. Mochizuki M, Furukawa N, Nagaosa N (2010) Theory of electromagnons in the multiferroic Mn perovskites: the vital role of higher harmonic components of the spiral spin order. Phys Rev Lett 104:177206
47. Fukuda T, Nojiri H, Motokawa M, Asano T, Mekata M, Ajiro Y (1998) Submillimeter wave ESR study on triangular lattice antiferromagnet $CuFeO_2$. Phys B 246:569
48. Moriya T (1968) Theory of absorption and scattering of light by magnetic crystals. J Appl Phys 39:1042
49. Kenzelmann M, Lawes G, Harris AB, Gasparovic G, Broholm C, Ramirez AP, Jorge GA, Jaime M, Park S, Huang Q, Shapiro AY, Demianets LA (2007) Direct transition from a disordered to a multiferroic phase on a triangular lattice. Phys Rev Lett 98:267205
50. Angelov S, Doumerc JP (1991) On the correlation between the structure and the exchange interactions in ACrO$_2$ chromites. Solid State Commun 77:213
51. Kadowaki H, Kikuchi H, Ajiro Y (1990) Neutron powder diffraction study of the two-dimensional triangular lattice antiferromagnet $CuCrO_2$. J Phys Condens Matter 2:4485
52. Kadowaki H, Takei H, Motoya K (1995) Double-Q 120 degrees structure in the Heisenberg antiferromagnet on rhombohedrally stacked triangular lattice $LiCrO_2$. J Phys Condens Matter 7:6869
53. Oohara Y, Mitsuda S, Yoshizawa H, Yaguchi N, Kuriyama H, Asano T, Mekata M (1994) Magnetic phase transition in $AgCrO_2$. J Phys Soc Jpn 63:847
54. Alexander LK, Büttgen N, Nath R, Mahajan AV, Loidl A (2007) ^{7}Li NMR studies on the triangular lattice system $LiCrO_2$. Phys Rev B 76:064429
55. Soubeyroux JL, Fruchart D, Delmas C, Flem GL (1979) Neutron powder diffraction studies of two-dimensional magnetic oxides. J Magn Magn Mater 14:159
56. Olariu A, Mendels P, Bert F, Ueland BG, Schiffer P, Berger RF, Cava RJ (2006) Unconventional dynamics in triangular heisenberg antiferromagnet $NaCrO_2$. Phys Rev Lett 97:167203
57. Yamasaki Y, Sagayama H, Goto T, Matsuura M, Hirota K, Arima T, Tokura Y (2007) Electric control of spin helicity in a magnetic ferroelectric. Phys Rev Lett 98:147204
58. Kimura K, Nakamura H, Ogushi K, Kimura T (2008) Magnetoelectric control of spin-chiral ferroelectric domains in a triangular lattice antiferromagnet. Phys Rev B 78:140401(R)
59. Kimura K, Nakamura H, Kimura S, Hagiwara M, Kimura T (2009) Tuning ferroelectric polarization reversal by electric and magnetic fields in $CuCrO_2$. Phys Rev Lett 103:107201
60. Soda M, Kimura K, Kimura T, Matsuura M, Hirota K (2009) Electric control of spin helicity in multiferroic triangular lattice antiferromagnet $CuCrO_2$ with proper-screw order. J Phys Soc Jpn 78:124703
61. Kimura K, Otani T, Nakamura H, Wakabayashi Y, Kimura T (2009) Lattice distortion coupled with magnetic ordering in a triangular lattice antiferromagnet $CuCrO_2$. J Phys Soc Jpn 78:113710
62. Soda M, Kimura K, Kimura T, Hirota K (2010) Domain rearrangement and spin-spiral-plane flop as sources of magnetoelectric effects in delafossite $CuCrO_2$. Phys Rev B 81:100406
63. Kan EJ, Xiang HJ, Zhang Y, Lee C, Whangbo M-H (2009) Density-functional analysis of spin exchange and ferroelectric polarization in $AgCrO_2$. Phys Rev B 80:104417
64. Singh K, Maignan A, Martin C, Simon Ch (2009) $AgCrS_2$: a spin driven ferroelectric. Chem Mater 21:5007
65. Hemberger J, Lunkenheimer P, Fichtl R, Krug von Nidda H-A, Tsurkan V, Loidl A (2005) Relaxor ferroelectricity and colossal magnetocapacitive coupling in ferromagnetic $CdCr_2S_4$. Nat (Lond) 434:364
66. Catalan G, Scott JF (2007) Relaxor ferroelectricity and colossal magnetocapacitive coupling in ferromagnetic $CdCr_2S_4$. Nat (Lond) 448:E4
67. Katsumata K (1994) Landolt-Bïrnstein, vol 3. Springer, Berlin, p 27J1

68. Sato T, Kadowaki H, Iio K (1995) Successive phase transitions in the hexagonal-layered Heisenberg antiferromagnets MnX_2 (X = Br, I). Phys B 213:224
69. Cable JW, Wilkinson MK, Wollan EO, Koehler WC (1962) Neutron diffraction investigation of the magnetic order in MnI_2. Phys Rev 125:1860
70. Kuindersma SR, Sanchez JP, Haas C (1981) Magnetic and structural investigations on NiI_2 and CoI_2. Phys B 111:231
71. Kadowaki H, Ubukoshi K, Hirakawa K, Martinez JL, Shirane G (1987) Experimental study of new type phase transition in triangular lattice antiferromagnet VCl_2. J Phys Soc Jpn 56:4027

Chapter 4
Magnetoelectric Response in $S = 1/2$ Chain Helimagnets

Introduction

Besides the triangular lattice antiferromagnet as investigated in Chap. 3, one-dimensional chain magnet with competing J_1 and J_2 is also known as a typical example of frustrated spin systems. In this chapter, we focus on the edge-shared CuO_2 (or $CuCl_2$) chain as an ideal stage to realize such a situation, and investigate their magnetoelectric response in detail.

Magnetism in Edge-Shared CuO_2 Chain

Hereafter, we assume the situation where each Cu^{2+} ion ($S = 1/2$) is on the center of oxygen square and forms edge-shared chains. The structure of edge-shared CuO_2 chain is illustrated in Fig. 4.1. Because the Cu-O-Cu bond angle is close to 90°, Kanamori-Goodenough rule suggests that the nearest neighbor exchange interaction (J_1) is ferromagnetic but relatively weak as compared with the antiferromagnetic next nearest neighbor interaction (J_2).

As already discussed in "Magnetically-Induced Ferroelectricity", the magnetic ground state in the J_1 versus J_2 model essentially depends on the ratio J_1/J_2. When we assume the classical Heisenberg spin, helimagnetic order is stabilized in the region $4 > |J_1/J_2|$. However, the magnetism in edge-shared CuO_2 chain compound is expected to reflect the quantum aspect of electron spin, since all attributes of $S = 1/2$ nature, low-dimensionality, and magnetic frustration contribute cooperatively to enhance quantum fluctuation. Recently, Furukawa et al. calculated the magnetic phase diagram as a function of (J_1/J_2) and exchange anisotropy Δ using the spin Hamiltonian

$$\mathcal{H} = \sum_{n=1,2} J_n \sum_j (S_j^x S_{j+n}^x + S_j^y S_{j+n}^y + \Delta S_j^z S_{j+n}^z), \tag{4.1}$$

S. Seki, *Magnetoelectric Response in Low-Dimensional Frustrated Spin Systems*, Springer Theses, DOI: 10.1007/978-4-431-54091-5_4,

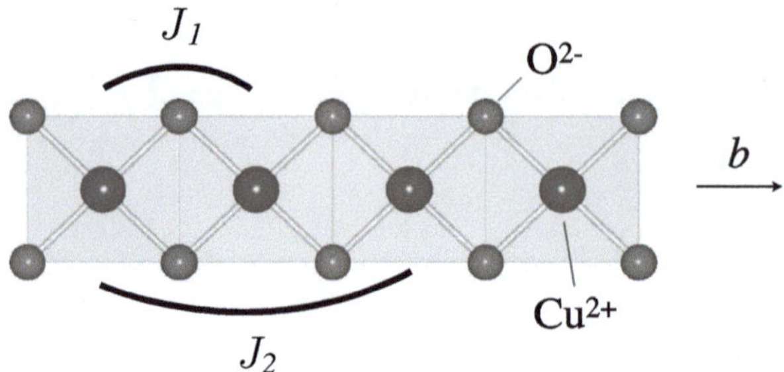

Fig. 4.1 The structure of edge-shared CuO_2 chain. The exchange paths for nearest neighbor interaction (J_1) and next nearest neighbor interaction (J_2) are also indicated

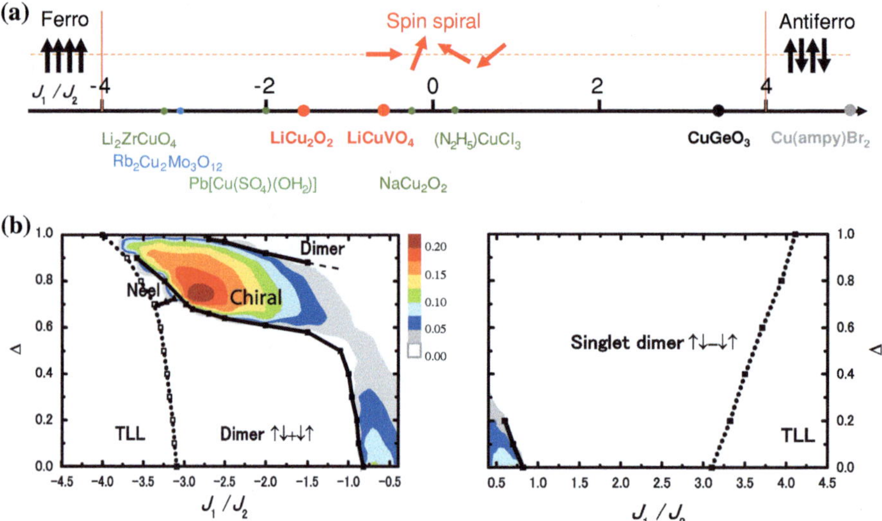

Fig. 4.2 **a** The classical phase diagram for J_1 versus J_2 model and relevant $S = 1/2$ chain magnets. **b** The quantum phase diagram calculated for J_1 versus J_2 model with spin-Hamiltonian as given by Eq. 4.1. (Adapted with permission from [1], ©2010 APS)

with the electronic spin $\mathbf{S}_j = (S_j^x, S_j^y, S_j^z)$ at the Cu^{2+} site j in the chain (Fig. 4.2) [1]. This phase diagram suggests the emergence of various types of non-trivial magnetic ground states such as singlet-dimer state. Still, the finite inter-chain coupling in the real material can partly suppress the effect of quantum fluctuation, which often stabilizes long-range helimagnetic order even in the $S = 1/2$ chain compounds.

Since 2007, the discovery of magnetically-induced ferroelectricity has been reported for two types of helimagnetic compounds with edge-shared CuO_2 chains; $LiCu_2O_2$ and $LiCuVO_4$. These compounds are reported to host the cycloidal spin order propagating along the chain direction, and are seemingly the ideal model

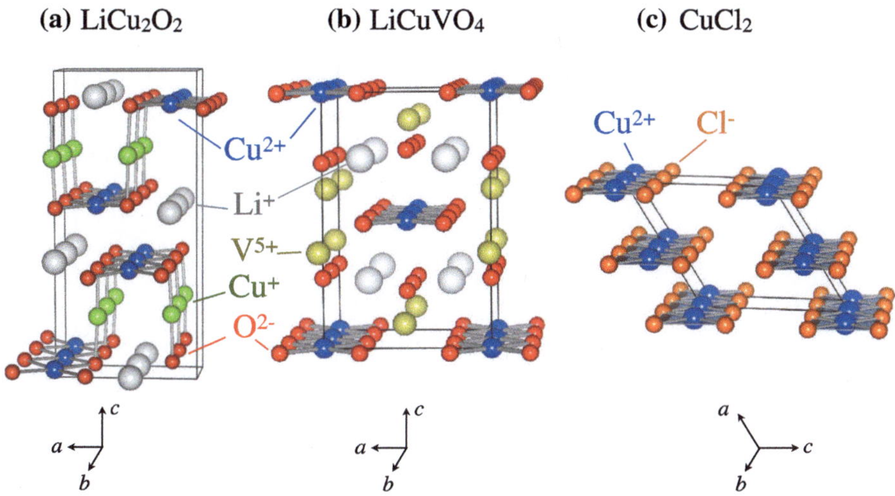

Fig. 4.3 Crystal structure of **a** $LiCu_2O_2$, **b** $LiCuVO_4$ and **c** $CuCl_2$, respectively

compounds to testify the prediction of the inverse D-M mechanism due to its simple lattice and magnetic geometry. However, the experimentally reported magnetoelectric response is not so straightforward, as mentioned below.

Magnetoelectric Response in $LiCu_2O_2$

$LiCu_2O_2$ takes the crystal structure as shown in Fig. 4.3a, which is characterized by the orthorhombic space group *Pnma* and lattice parameters $a = 5.73$, $b = 2.86$ and $c = 12.4$ Å [3]. This material contains equal number of Cu^{1+} and Cu^{2+}, and only the latter of which carries spin $S = 1/2$ and forms edge-shared chain structure running along the b-axis with CuO_2 square plaquettes lying in the ab-plane. The magnitude of inter-chain interaction is presumed to be small ($<|J_1|, |J_2|$), though has not reached the consensus as yet [4, 5]. As a result of the frustration, a spiral magnetic structure is realized below $T_{N2} \sim 23$ K. A former (unpolarized) neutron diffraction study has revealed the incommensurate magnetic structure with the modulation vector (0.5, 0.174, 0), and claimed the cycloidal spin order with ab-spiral spin plane (Fig. 4.4b) [3]. In this phase, however, the appearance of spontaneous electric polarization along the c-axis has recently been reported [2]. To reconcile the observed polarization direction with the spiral spin state, the inverse D-M model requires the bc-cycloidal spin order(Fig. 4.4a), not the reported ab-cycloid. $LiCu_2O_2$ also shows the P-flop transition from $P \parallel c$ to $P \parallel a$ under $H \parallel b$ (Fig. 4.4c) [2]. If we assume the bc-cycloidal spin order to reproduce $P \parallel c$ at 0 T, $H \parallel b$ is expected to stabilize the proper screw spin order with ac-spin spiral plane. Since the latter spin texture is paraelectric according to the inverse D-M model, the emergence of $P \parallel a$

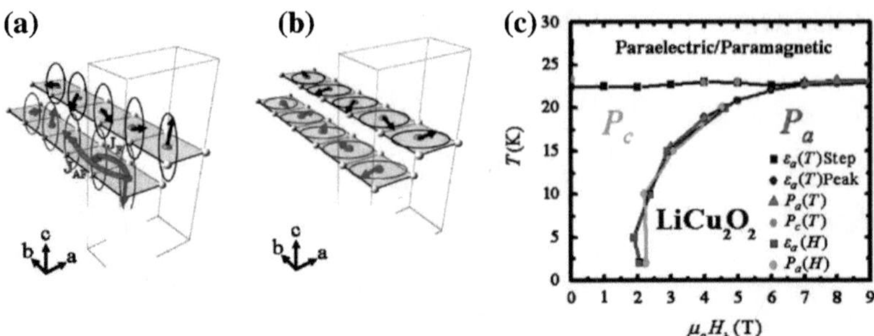

Fig. 4.4 **a** *bc*-cycloidal spin order and **b** *ab*-cycloidal spin order on $LiCu_2O_2$. **c** $H - T$ magnetoelectric phase diagram for $LiCu_2O_2$ under $H \parallel b$. (Adapted with permission from [2], ©2007 APS)

under $H \parallel b$ is also seemingly strange. Several recent diffraction studies suggested a more complex spin order rather than simple cycloid may be realized [1, 6, 7], and the magnetic structure of the ferroelectric ground state is still under controversy. Notably, the powder neutron scattering study on the isostructural material $NaCu_2O_2$ justifies the *bc*-spiral spin structure, and the magnetic moment of Cu^{2+} is estimated as small as $0.56\,\mu_B$ [8]. This implies that the effect of quantum fluctuation is important also in $LiCu_2O_2$.

Magnetoelectric Response in $LiCuVO_4$

$LiCuVO_4$ takes the crystal structure as shown in Fig. 4.3b, which is characterized by the orthorhombic space group *Imma* and lattice parameters $a = 5.66$, $b = 5.81$ and $c = 8.76$ Å. As in case of $LiCu_2O_2$, this compound possesses edge-shared chain structure running along the *b*-axis with CuO_2 square plaquettes lying in the *ab*-plane. Previous neutron diffraction study on single crystal suggested that the *ab*-cycloidal spin order with magnetic modulation vector (0, 0.53, 0) is realized below 2.1 K [9]. This helimagnetic ground state was found to induced $P \parallel a$ [11], which is consistent with the prediction of the inverse D-M model. When magnetic field is applied parallel to the *a*-axis, this compound shows spin-flop transition into *bc*-cycloidal spin state. While the inverse D-M model predicts that such a spin-flop transition should lead to the switching of *P*-direction from $P \parallel a$ to $P \parallel c$, the latter $P \parallel c$ component cannot be detected in the measurement by Yasui et al. [12]. Contrastingly, Schrettle et al. reported the observation $P \parallel c$ under $H \parallel a$ (Fig. 4.5) [10]. At this stage, the origin of this discrepancy is not clear.

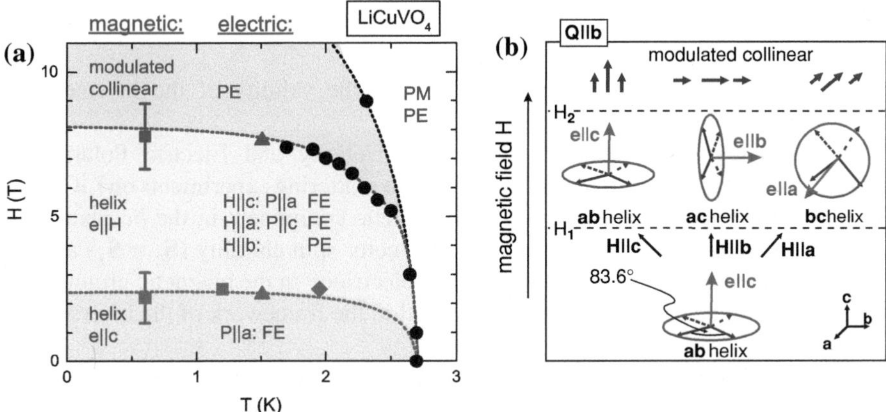

Fig. 4.5 **a** $H-T$ magnetoelectric phase diagram and **b** Evolution of spin-orientation under various directions and strength of magnetic field, determined for $LiCuVO_4$ by Schrettle et al. (Adapted with permission from [10], ©2008 APS)

Origin of Magnetically-Induced Ferroelectricity in $S = 1/2$ *Chain Helimagnets*

As discussed above, several conflicting experimental results are reported for both $LiCu_2O_2$ and $LiCuVO_4$, and the validity of the inverse D-M model for these $S = 1/2$ chain helimagnets is still in controversy. Notably, both compounds are reported to frequently contain the Li-Cu intersubstitution due to their close ionic radii. The reported chemical composition of as-grown $LiCu_2O_2$ single crystal is $Li_{1.16}Cu_{1.84}O_{2.01}$, where 16 % of Cu^{2+} sites are replaced with Li^+ ion and 16 % of Cu^+ sites are done by Cu^{2+} [3]. Moskvin et al. suggested that such crystallographic defects and exchange striction are the origin of ferroelectricity for $LiCu_2O_2$ [13, 14] and $LiCuVO_4$ [15]; In fact, their calculation seems to reproduce several anomalous magnetoelectric behavior such as the emergence of $P \parallel c$ in the ab-cycloid spin state for $LiCu_2O_2$ as well as the transition from $P \parallel a$ to $P = 0$ under $H \parallel a$ for $LiCuVO_4$. However, some other groups believe that the inverse D-M mechanism is still active in these compounds, and suggested that observed anomalous magnetoelectric response may come from the complexity of spin structure [2, 16]. Since it is also not clear how the strong quantum fluctuation can affect the magnetoelectric response in the frustrated magnets, the elucidation of the magnetoelectric coupling mechanism in the $S = 1/2$ chain magnets will be important.

Overview

Our main purpose in this chapter is to check the validity of the inverse D-M mechanism in the $S = 1/2$ chain helimagnets.

In "$LiCu_2O_2$: Correlation Between Spin-Helicity and Electric Polarization Vector", we report the result of polarized neutron scattering experiments on $LiCu_2O_2$. We have confirmed the existence of helimagnetic component in the bc-plane, and also demonstrated the coupling between the vector spin chirality ($\mathbf{S}_i \times \mathbf{S}_j$) and the sign of P. These results suggest that the ferroelectricity in the magnetic ground state of $LiCu_2O_2$ can be consistently explained within the framework of the inverse D-M model.

In "$CuCl_2$", we investigate the magnetoelectric response in $CuCl_2$, which is also characterized by the $S = 1/2$ edge-shared chain structure and cycloidal spin order. Because the ionic radius of Cu^{2+} and Cl^- is quite different, $CuCl_2$ should be free from the chemical intersubstitution as observed in the Li-based compounds. We discovered the emergence of magnetically-induced ferroelectricity in $CuCl_2$, and the behavior of P under various direction of H well reproduces the prediction of inverse D-M model. The observed magnetoelectric response for $CuCl_2$ is quite similar to that reported for $LiCuVO_4$ by Schrettle et al. [10], which suggests that the conflicting results reported for $LiCuVO_4$ by Yasui et al. [12] may be caused by the crystallographic imperfection in their specimen.

$LiCu_2O_2$: Correlation Between Spin-Helicity and Electric Polarization Vector

Introduction

In this section, to clarify the origin of ferroelectricity in $LiCu_2O_2$, we testify the validity of the inverse D-M model for the e_g-electron spin system with potentially large quantum fluctuation. According to Eq. (1.10), the inverse D-M model predicts the coupling between the sign of P-direction and vector spin chirality ($\mathbf{S}_i \times \mathbf{S}_j$). Recent polarized neutron scattering study on $TbMnO_3$ has experimentally confirmed such polarity-dependent vector chirality [17], which is now considered as the definitive evidence for the spiral-spin driven ferroelectricity. We performed the related experiments on $LiCu_2O_2$, and also reinvestigated the orientation of spin-spiral plane to resolve the reported contradiction between spin structure and induced P-direction.

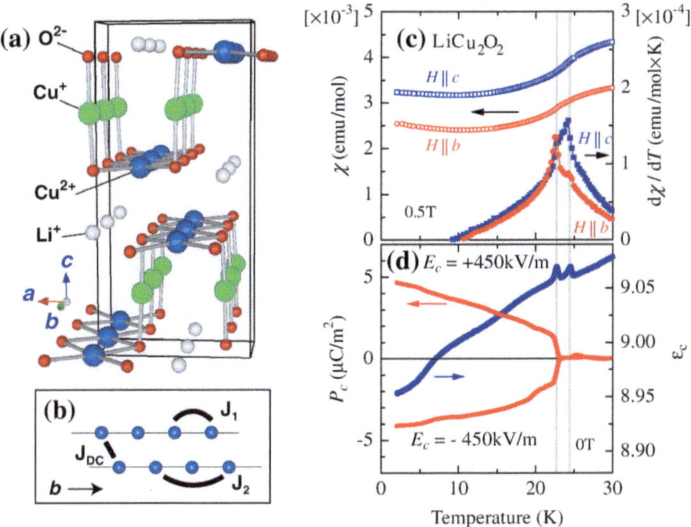

Fig. 4.6 **a** Crystal structure of $LiCu_2O_2$. **b** Schematic view of magnetic interactions between Cu^{2+} sites. **c**, **d** Temperature dependence of magnetic susceptibility, electric polarization and dielectric constant. All the quantities were measured in the warming process

Results

Neutron diffraction is known as one of most powerful experimental methods to determine the magnetic structure in solids. The employment of polarized neutron provides additional sensitivity to the spin-orientation and spin-helicity (clockwise or counterclockwise manner of spin rotation) of the target compound, which enables more precise characterization of non-collinear magnetic structure [18]. The detailed setup of the polarized neutron scattering experiment is described in "Polarized Neutron Scattering".

Figure 4.6c and d show the temperature dependence of magnetic susceptibility, dielectric constant, and electric polarization for $LiCu_2O_2$. For $H \parallel c$, the temperature derivative of magnetic susceptibility ($d\chi/dT$) indicates two anomalies at $T_{N1} \sim 24.5$ and $T_{N2} \sim 23.0$ K, although only one peak at T_{N2} is found in $d\chi/dT$ for $H \parallel b$ (or a). These imply the existence of two magnetic phases below T_{N1}; AF1 ($T_{N1} > T > T_{N2}$) and AF2 ($T_{N2} > T$). The anomaly at 9K possibly caused by impurity Li_2CuO_2 [3, 19] was absent in our sample. The spontaneous electric polarization parallel to the c-axis (P_c) evolves only below T_{N2}. The P_c can be reversed with the opposite poling electric field (E_c). This indicates the ferroelectric nature of AF2 phase, and suggests the correlation between ferroelectricity and magnetic properties. All these features reproduced the results reported by Park et al. [2, 6], who proposed the sinusoidal spin structure with collinear spins (parallel to the c-axis) for AF1. A recent theory proposed the intriguing scenario of the novel cholesteric spin state for this phase [20].

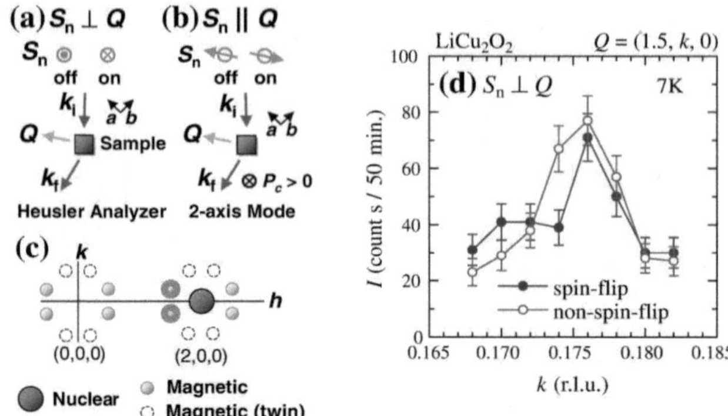

Fig. 4.7 The experimental geometries for the polarized neutron diffraction; **a** $\mathbf{S}_n \perp \mathbf{Q}$ and **b** $\mathbf{S}_n \parallel \mathbf{Q}$. The labels "*on*" and "*off*" indicate the state of the neutron-spin flipper. **c** Schematic illustration of nuclear and magnetic Bragg positions in the reciprocal space. **d** The k-scan profiles of the $(1.5, +\delta, 0)$ magnetic reflection in the $\mathbf{S}_n \perp \mathbf{Q}$ setup

We measured the poling electric field dependence of spontaneous polarization and confirmed that the saturation of P_c was achieved above $|E_c| \sim 350\,\text{kV/m}$. We also measured dielectric constant parallel to the c-axis (ϵ_c) and found peaks at both T_{N1} and T_{N2}, although previously only one peak at T_{N2} was reported [2].

For the polarized neutron diffraction measurements, we focused on the ferroelectric AF2 phase. Since different magnetic structures, such as the ab-spiral [3] and the bc-spiral plus a-component structure [2, 6], have been proposed for this phase, whether the magnetic moment is present along the c-axis was first examined. For this purpose, we took the $\mathbf{S}_n \perp \mathbf{Q}$ setup (Fig. 4.7a), where neutron spins were parallel or antiparallel to the c-axis. To distinguish between the spin-flip and non-spin-flip scattering, a Heusler analyzer was employed. In general, only the magnetic moment perpendicular to $\mathbf{Q}$ contributes to the magnetic reflection of neutrons. For polarized neutrons, furthermore, the magnetic moment parallel to $\mathbf{S}_n$ produces the non-spin-flip scattering and the moment perpendicular to $\mathbf{S}_n$ does the spin-flip scattering [21]. Figure 4.7d shows the k-scan profile of the $(1.5, +\delta, 0)$ magnetic reflection at 7 K ($< T_{N2}$). The observed modulation wavenumber, $\delta \sim 0.175$, is in accord with literature [3]. Since $\mathbf{Q}$ can be considered almost parallel to the a-axis in this configuration (Fig. 4.7c), the b-component of magnetic moment (m_b) contributes to the spin-flip scattering while the c-component (m_c) to the non-spin-flip scattering. Assuming the common background for the both profiles, the integrated intensities are nearly equal (spin-flip(m_b)/non-spin-flip(m_c) ≈ 0.9). This suggests the existence of the nearly same weight of b- and c-components in the magnetic structure of AF2. This is consistent with the bc-spiral (or plus some a-component) model [2, 6], and at least not with the simple ab-spiral one [3].

Next, we attempted to observe the relationship between the polarization direction and the chirality of spin spiral. For this purpose, we adopted the $\mathbf{S}_n \parallel \mathbf{Q}$ setup (Fig. 4.7b), where neutron spins are parallel or antiparallel to $\mathbf{Q}$. In this

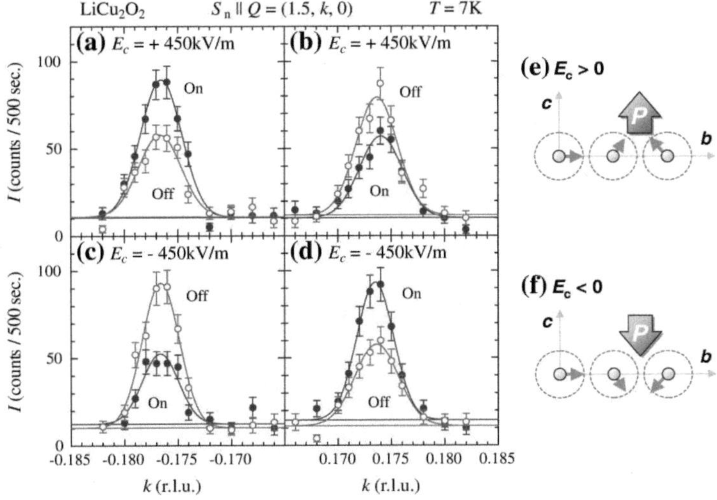

Fig. 4.8 **a–d** The k-scan profiles of the $(1.5, \pm\delta, 0)$ magnetic reflections in the $\mathbf{S}_n \| \mathbf{Q}$ setup. The labels "*on*" and "*off*" show the state of neutron-spin flipper. Solid lines show the result of the Gaussian fitting. **e**, **f** The geometrical relationships between spin chirality (helicity) and electric polarization determined from the observed results

alignment, only spin-flip scatterings contribute to the magnetic reflection. Therefore, no
polarization analysis is needed, and we employed the two-axes mode without an analyzer. Figure 4.8a–d show the k-scan profiles of the $(1.5, \pm\delta, 0)$ magnetic reflection at 7 K[1] with various poling electric fields parallel to the c-axis (E_c). E_c was applied at 30 K ($>T_{N1}$) and removed at 7 K just before the diffraction measurements to obtain a single ferroelectric domain. With $|E_c| = 450$ kV/m, the difference of intensity between $\pm\delta$ was clearly observed, and the relative intensity was confirmed to be reversed by changing the sign of either $\mathbf{S}_n$ or E_c. These behaviors can be interpreted in terms of the E_c-dependent vector chirality of the transverse bc-spiral spins as follows.

According to Blume [18], the magnetic cross section for polarized neutron is given as

$$\left(\frac{d\sigma}{d\Omega}\right) \propto \sum_{i,j} \exp\{i\mathbf{Q}(\mathbf{R}_i - \mathbf{R}_j)\}\left[\eta_j \cdot \eta_i + i\hat{\mathbf{S}}_n(\eta_j \times \eta_i)\right] \tag{4.2}$$

Here, η_i denotes the component of $\mathbf{m}_i$ perpendicular to $\mathbf{Q}$, $\eta_i = \hat{\mathbf{Q}} \times (\mathbf{m}_i \times \hat{\mathbf{Q}})$, where $\hat{\mathbf{Q}} = \mathbf{Q}/|\mathbf{Q}|$ and $\hat{\mathbf{S}}_n = \mathbf{S}_n/|\mathbf{S}_n|$. For simplicity, we take hereafter the approximation that $\mathbf{S}_n \| \mathbf{Q} \| a$ and define $abc \rightarrow zxy$, where z is the spin quantization axis. Then, the spin vector chirality on the bc-plane can be defined as $\mathbf{C} = (\eta_i \times \eta_j)/|\eta_i \times \eta_j|$. With use of the relations $\eta_i = (\sigma_i^x, \sigma_i^y, 0)$ and $\sigma^{\pm} = \sigma^x \pm i\sigma^y$, the cross section for

[1] Slightly different wavenumbers between $\pm\delta$ is perhaps due to the misalignment of the sample.

the $(1.5, \pm\delta, 0)$ magnetic reflections can be expressed as

$$\left(\frac{d\sigma}{d\Omega}\right)_{\pm} = \left(\frac{d\sigma}{d\Omega}\right)_{c} \pm \left(\frac{d\sigma}{d\Omega}\right)_{s} \tag{4.3}$$

where

$$\left(\frac{d\sigma}{d\Omega}\right)_{c} \propto \sum_{i,j} \cos\{\mathbf{Q}(\mathbf{R}_i - \mathbf{R}_j)\} \cdot \langle \sigma_i^{+} \sigma_j^{-} \rangle \tag{4.4}$$

$$\left(\frac{d\sigma}{d\Omega}\right)_{s} \propto \sum_{i,j} \sin\{\mathbf{Q}(\mathbf{R}_i - \mathbf{R}_j)\} \cdot \langle \sigma_i^{+} \sigma_j^{-} \rangle \tag{4.5}$$

For intuitive understanding, we tentatively treat the cross section in the classical limit. Based on the results for the $\mathbf{S}_{\mathrm{n}} \perp \mathbf{Q}$ setup, we can assume the bc-spiral magnetic structure plus several a-component:

$$\begin{aligned} \mathbf{m}_i = & \, m_b \cdot \mathbf{e}_b \cdot \cos(\mathbf{q}_m \mathbf{R}_i) + m_c \cdot \mathbf{e}_c \cdot \sin(\mathbf{q}_m \mathbf{R}_i) \\ & + m_a \cdot \mathbf{e}_a \cdot \sin(\mathbf{q}_m \mathbf{R}_i + \delta') \end{aligned} \tag{4.6}$$

Here, $\mathbf{e}_a$, $\mathbf{e}_b$, and $\mathbf{e}_c$ are the unit vectors along the a, b, and c-axis. Then, Eq. (4.3) can be written as [17, 18]

$$\left(\frac{d\sigma}{d\Omega}\right)_{\pm} \propto \left[m_b^2 + m_c^2 \pm 2m_b \cdot m_c \cdot (\hat{\mathbf{S}}_{\mathrm{n}} \cdot \hat{\mathbf{Q}})(\hat{\mathbf{Q}} \cdot \mathbf{C}) \right] \tag{4.7}$$

The last term predicts the different scattering intensities for $\pm\delta$, and the relation can be reversed by changing the sign of either $\mathbf{S}_{\mathrm{n}}$ or $\mathbf{C}$. In fact, this behavior is clearly observed in the results with $E_c = +450$ kV/m (Fig. 4.8a and b). This means that $\hat{\mathbf{Q}} \cdot \mathbf{C}$ is not zero, or in other words the magnetic structure of AF2 has the spiral components in the bc-plane. Moreover, when the sign of E_c is reversed, the differential intensity relation is also reversed (Fig. 4.8c and d). This indicates that the spin chirality determines the direction of electric polarization. Conversely, the observed electric control of spin helicity directly proves that the ferroelectricity of $LiCu_2O_2$ originates from the transverse-spiral (cycloidal) spin structure. Thus, the inverse D-M model holds good even for the e_g-electron spin system, or under possibly large quantum fluctuation inherent to the frustrated $S = 1/2$ spins. The obtained geometric relation between spin chirality and electric polarization is illustrated in Fig. 4.8e and f. The sign of the coupling constant in Eq. (1.10) is negative ($A < 0$), which agrees with the theoretical prediction [22]. Note that the sign of A[2] is different from the case of $TbMnO_3$ [17]. We also measured the profiles with $E_c = 0$ and found no difference for the intensity between $\pm\delta$ reflections nor between the neutron spin

---

[2] Sign of the $\mathbf{S}_{\mathrm{n}}$-dependent term in Eq. (4.7) is wrong in Ref. [17], because of the different definition of $\mathbf{Q}$ from Ref. [18]. In the correct definition, $A > 0$ is obtained for $TbMnO_3$.

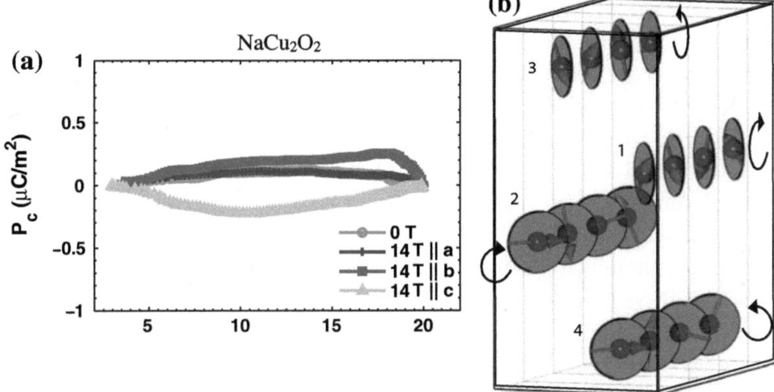

Fig. 4.9 **a** Temperature dependence of electric polarization P for $NaCu_2O_2$. **b** Antiferro-chiral magnetic ground state for $NaCu_2O_2$ determined by the neutron diffraction study on single crystal. (Adapted with permission from [23, 24], ©2010 APS)

states. This should be due to the coexistence of opposite ferroelectric domains (or clockwise/counter-clockwise spin-spiral domains) for the zero electric-field case.

Discussion

An unresolved problem at this stage is the ratio of scattering intensity between the stronger and weaker reflections. From Eq. (4.7), the elliptic ratio of the spiral spin, m_b/m_c (or m_c/m_b), is estimated as $|(\sqrt{I_{\mathrm{ON}}} - \sqrt{I_{\mathrm{OFF}}})/(\sqrt{I_{\mathrm{ON}}} + \sqrt{I_{\mathrm{OFF}}})|$ for the case of classical spin [17]. On the basis of the data shown in Fig. 4.8a–d, this expression gives m_c/m_b (or m_b/m_c) $= 0.09 \sim 0.20$. On the other hand, the afore-mentioned results on the $\mathbf{S}_{\mathrm{n}} \perp \mathbf{Q}$ setup suggests the nearly equal value for m_b and m_c. As the origin of this discrepancy, the coexistence of different polarity domains might be suspected. However, we confirmed the saturation of electric polarization at $|E_c| = 350\,\mathrm{kV/m}$, with the same (Al) electrode used in the neutron scattering study. Also on the same sample, the Ag electrode was tested to confirm the identical saturation value of electric polarization. Therefore, we believe that the single domain state was realized in the $\mathbf{S}_{\mathrm{n}} \| \mathbf{Q}$ setup, and the above apparent discrepancy should be ascribed to a more intrinsic origin. The measured temperature (7 K) might not be low enough to saturate the spin order. However, the P value at 7 K already reaches 80–90 % of the 2 K value (see Fig.4.6d); thermal fluctuation alone is not enough to decrease the spin ellipticity m_c/m_b.

Recently, several groups have suggested that the magnetic structure of AF2 would be more complicated than the simple bc-spiral [1, 2, 6, 7]. Such complex magnetic order must require some modification in Eq. (4.7). Notably, recent single-crystal neutron diffraction study by Capogna et al. concluded that the isostructural $NaCu_2O_2$ hosts the antiferro-chiral spin ordering between two adjacent chains [23], which leads to the antiferroelectric state accompanied with the absence of electric polarization (Fig. 4.9) [24]. Since $NaCu_2O_2$ is confirmed to be free from the crystallographic defects unlike the case of $LiCu_2O_2$, the intense Li-Cu intersubstitution in $LiCu_2O_2$ may cause the small imbalance between two opposite spin-chirality and allow the emergence of finite P. The seemingly strange P-flop behavior under $H \parallel b$ may also arise from such complexity of magnetic order. Note however that even with any other magnetic structure the observed difference for the opposite neutron spins $\mathbf{S}_n$ reflects the chirality in the bc-plane (see Eq. (4.2)), thus supports the validity of the inverse D-M mechanism.

One of other possibilities is the effect of quantum fluctuation. In case of $S = 1/2$ quantum-spin systems like $LiCu_2O_2$, the validity of the classical-spin treatment as done in Eqs. (4.6) and (4.7) is no longer guaranteed. For a more rigorous argument, we have to go back to Eqs. (4.3–4.5). According to these expressions, both $(d\sigma/d\Omega)_c$ and $(d\sigma/d\Omega)_s$ are the Fourier components (symmetric and antisymmetric, respectively) of the same physical quantity $\sigma_i^+\sigma_j^-$. Therefore, the distribution of scattering intensities reflects the balance between symmetric and antisymmetric components of $\sigma_i^+\sigma_j^-$ for the $S = 1/2$ case. This may be the cause of the deviation from the Eq. (4.7). For example, in the extreme case of quantum fluctuation where the spins form the singlet state, the commutation that $\langle\sigma_i^+\sigma_j^-\rangle = \langle\sigma_j^+\sigma_i^-\rangle$ holds, therefore $(d\sigma/d\Omega)_s = 0$ and no differential intensity should be observed. The experimental observation of shrunk magnetic moment [8] implies the large quantum fluctuation subsisting in the ordered spiral state. Therefore, the quantum fluctuation of the vector spin chirality is likely to result in the reduced differential $\pm\delta$ reflection intensity of polarized neutrons, as observed. Several latest theories indeed reproduced the observed scattering profiles considering the effect of quantum fluctuation [25, 26]. For the thorough understanding, further analysis of the magnetic structure and its quantum dynamics will be needed.

Conclusion

In this section, the polarized neutron diffraction study was performed on the quantum-spin chain magnet $LiCu_2O_2$. We confirmed the coupling between spin vector chirality of the transverse bc-spiral structure and the direction of electric polarization along the c-axis. This proves that even with the e_g-electron system under the large quantum fluctuation the spin-current model or the inverse Dzyaloshinskii-Moriya mechanism still works. The differential intensity of polarized neutron reflections show a clear deviation from that expected for the classical bc-spiral spin structure, which may reflect the complexity of magnetic order and/or the effect of quantum fluctuation in this $S = 1/2$ chain compound.

$CuCl_2$

Introduction

In this section, we investigate the magnetoelectric response in $CuCl_2$, which also possesses $S = 1/2$ edge-shared chain structure and cycloidal spin order as in the case for $LiCu_2O_2$ and $LiCuVO_4$. While frequent Li-Cu intersubstituion in Li-based compounds seems to cause the conflicting experimental results and prevent the appropreate interpretation of observed magnetoelectric response, $CuCl_2$ is expected to be free from such crystallographic defects due to the large difference of the ionic radius between Cu^{2+} and Cl^-.

Anhydrous cupric chloride $CuCl_2$ crystalizes into distorted CdI_2 form with monoclinic $C2/m$ space group and $\beta = 122°$ [28]. While original CdI_2 structure consists of the stacking of triangular lattices along the z-axis,[3] they are extended along the a-axis due to Jahn-Teller active Cu^{2+} ions (Fig. 4.10a). As a result, $CuCl_2$ can be regarded as the aggregate of edge-shared chains running along the b-axis, with $CuCl_4$ square plaquettes lying in the bc-plane (Fig. 4.10b). Magnetism is dominated by the intra-chain coupling between Cu^{2+} ($S = 1/2$) ions, and competition between ferromagnetic nearest-neighbor interaction and antiferromagnetic next-nearest-neighbor interaction stabilizes the helimagnetic ground state below 24 K [27, 29, 30]. Recent powder neutron scattering study suggested the cycloidal magnetic order propagating along the b-axis, with spin spiral confined in the bc-plane (Fig. 4.10b) and propagation vector $q \sim (1, 0.226, 0.5)$ [27]. While no dielectric measurements have been reported, the latest calculation based on density functional theory (DFT) predicts emergence of ferroelectricity along the c-axis [27]. In the following, we report the experimental discovery of FE and ME natures in $CuCl_2$, and prove that the IDM mechanism is still robust even under the strong quantum fluctuation. $CuCl_2$ is also among the first example of non-chalcogen based spiral-spin induced multiferroics, which promises further discovery of unique magnetoelectric function in many MX_2-type compounds and other forms of halide compounds.[4]

Results

First, we performed the characterization of our single crystal specimen. As suggested in [27], the ac-twin domains are expected to readily occur in $CuCl_2$. To check this possibility, we first performed ESR measurements under various directions of H confined within the ac-plane (Fig. 4.10d). Hereafter, we define θ_H as the angle be-

[3] For simplicity, hereafter we define the z-axis as the direction perpendicular to both a- and b-axes (Fig. 4.10c).

[4] For the magnetoelectric response of other MX_2-type halides, see "MX_2-Type Halides with CdI_2 Structure".

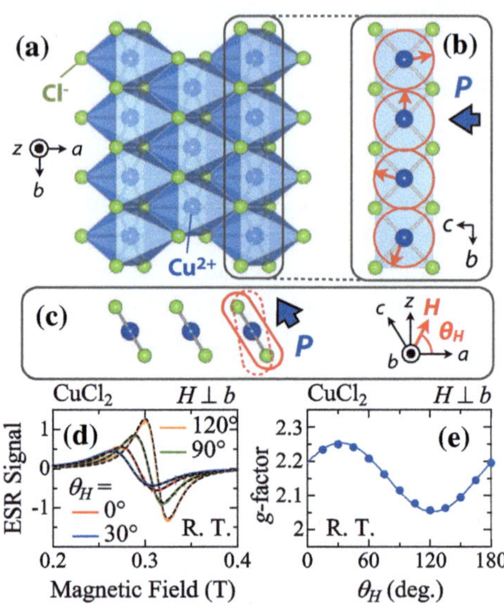

Fig. 4.10 **a–c** Crystal structure of $CuCl_2$, and P-direction observed at the ground state. The bc-cycloidal spin order suggested by Banks et al. [27] is illustrated in **(b)**, and also in **(c)** with solid rounded square representing spin-spiral plane. *Dashed rounded square* indicates the possible tilting of spin-spiral plane as revealed in this study (see text). **d** ESR signal taken at room temperature under various directions of H confined within the ac-plane. θ_H is defined as the angle between the a-axis and H-direction. Each *dashed line* represents a fitted curve with a single Lorentzian resonance. **e** Angle dependence of g-factor

tween the a-axis and H-direction. Each observed profile can be fitted well with a single Lorentzian resonance for all θ_H, indicating our crystal grown by the Bridgman method has no crystallographic twinning. The deduced g-factor shows sinusoidal θ_H-dependence (Fig. 4.10e), whose maximum and minimum values agree well with those previously reported for a twinned crystal [27].

Next, we measured T-dependence of magnetic susceptibility χ, ϵ, and P (Figs. 4.11a–c). χ suddenly drops at $T_N \sim 24$ K, which signals the transition into a spiral magnetic phase. Simultaneously, z-component of ϵ (ϵ_z) shows a sharp anomaly and a- and z-components of P (P_a and P_z) begin to develop. P-direction was reversed for opposite sign of poling E, and no P_b component could be confirmed. These results imply strong correlation between helimagnetic and FE orders in $CuCl_2$. Based on the bc-plane helimagnetic structure suggested in [27], the IDM model as well as the DFT calculation [27] predicts $|P_a/P_z| \sim 0.64$ (i.e. $P \parallel c$). This roughly agrees with the observed $|P_a/P_z| \sim 0.70$.

Figure 4.11d and e indicate H-dependence of M, ϵ, and P for $H \parallel b$. At 4 T, M-profile shows a clear step as already reported [27, 31]. Concomitantly, ϵ_z shows a sharp peak and both P_a and P_z are drastically suppressed. Since antiferromagnetic spins favor to lie within a plane perpendicular to H, this transition should correspond to a spin-flop into the ac-spiral spin state. The ac-spiral spin structure belongs to a magnetic form called proper screw, where spin-spiral plane is perpendicular to the modulation vector along the b-axis. The IDM model predicts $P = 0$ for this type of spin order due to the relationship $(\vec{S}_i \times \vec{S}_j) \parallel \vec{e}_{ij} \parallel b$, which is consistent with the observed suppression of P. Figure 4.11f summarizes the obtained $H - T$ phase

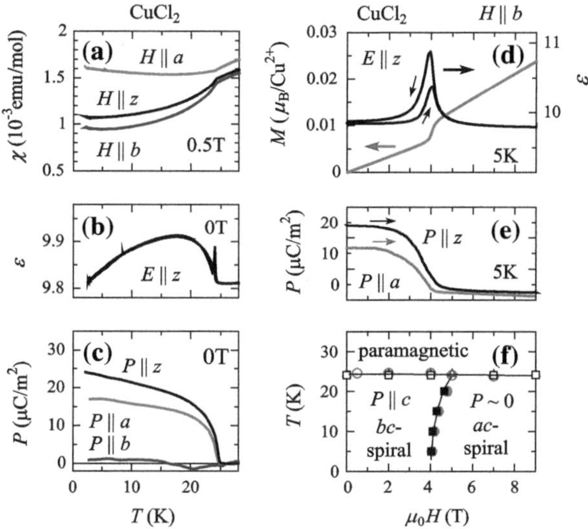

Fig. 4.11 Temperature dependence of **a** magnetic susceptibility χ, **b** dielectric constant ϵ, and **c** electric polarization P. In **d** and **e**, H-dependences of magnetization M, ϵ, and P under $H \parallel b$ are indicated. *Large* and *small arrows* denote corresponding ordinate scale of physical quantity and direction of field scans, respectively. **f** $H-T$ phase diagram for $H \parallel b$. *Circles*, *squares*, *triangles* are the data points obtained from M, ϵ and P profiles, respectively. *Open* (*closed*) *symbols* are taken from T- (H-) increasing runs

diagram for $H \parallel b$. The boundary of the FE phase always coincides with that for magnetic phases, which proves the interplay between FE and helimagnetic natures.

We further investigated the properties under $H \perp b$. Here, we adopt the same definition of θ_H as used for ESR measurements. Figure 4.12a indicates H-dependence of M measured at various θ_H. While no magnetic transition has been reported for $H \perp b$ [31], we found a clear signature of spin-flop at $H_{SF} \sim 4$ T most pronounced around $\theta_H = 100°$. θ_H-dependence of $\chi (= M/H)$ was also measured (Fig. 4.12b), and χ sinusoidally changes with minimum at $\theta_H \sim 100°$ below H_{SF}. In general, the sharpest transition of spin-flop as well as the minimum value of χ should be observed when H is applied parallel to the magnetic easy-plane. These results imply the magnetic easy-plane, i.e. spin-spiral plane at the ground state, is tilted from the originally suggested bc-plane toward the bz-plane by about 20° (Fig. 4.10c). Above H_{SF}, χ still modulates sinusoidally but with different χ-minimum position at $\theta_H \sim 122°$ (i.e. $H \parallel c$). With $H > H_{SF}$, the gain of Zeeman energy exceeds the energy loss due to magnetic anisotropy, and continuous rotation of spin-spiral plane is expected. In this case, θ_H-dependence of χ rather reflects the anisotropy of g-value [32], whose minimum is also confirmed to appear at $H \parallel c$ (Fig. 4.10e).

To investigate the behavior of P under H rotating around the b-axis, we simultaneously measured P_z and P_a using two pairs of electrodes. Thus, both P and H can be expressed as vectors within the ac-plane. We also define θ_P as the angle be-

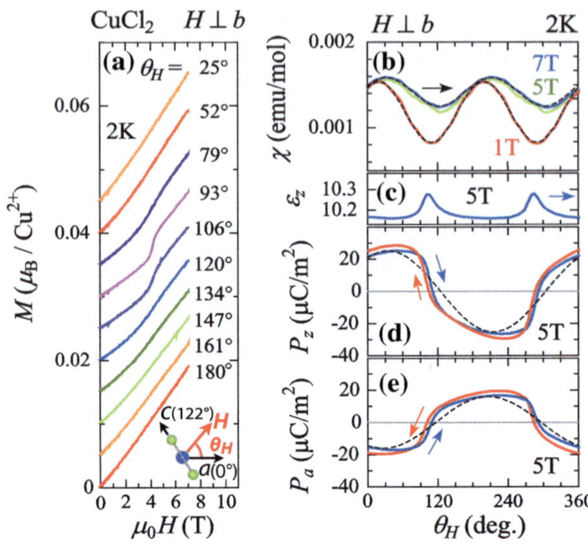

Fig. 4.12 **a** H-dependence of M taken under various directions of H around the b-axis. The base lines of data are arbitrarily shifted. **b** Angle-dependence of χ measured under $H \perp b$. *Dashed lines* indicate the fits with sinusoidal function. **c–e** Corresponding development of z-component of ϵ (ϵ_z) as well as z- and a-components of P (P_z and P_a). Before measurements of P, the specimen was cooled at $\theta_H = 0$ with poling E applied along the z-axis. *Dashed lines* indicate the behavior expected from Eq. (4.8). *Arrows* indicate the direction of H-rotation

tween the a-axis and observed P-direction (Fig. 4.13d). Figure 4.12d and e indicate θ_H-dependences of P_z and P_a measured at 5 T. When H is rotated by 180°, P-direction is always found to be reversed. To see the behavior of P more straightforwardly, the trace of P is plotted in the $P_a - P_z$ plane (Fig. 4.13a). It forms a shape like elongated ellipse. In Fig. 4.13b and c, θ_H-dependences of $|P|$ (magnitude of $\vec{P}$) and θ_P are indicated. θ_P takes almost constant value around $\theta_P = 120$ or 300°, suggesting the major axis of observed P-ellipse is pointing at the c-axis. If we assume that H is always perpendicular to the spin-spiral plane, i.e. $\vec{H} \parallel (\vec{S}_i \times \vec{S}_j)$, the IDM model predicts $\vec{P} = \vec{P}_1 \parallel c$ for $\vec{H} \perp c$ (Fig. 4.13e) and $\vec{P} = \vec{P}_2 \perp c$ for $\vec{H} \parallel c$ (Fig. 4.13f). For general θ_H, $\vec{P}$ is given as

$$\vec{P} = \vec{P}_1 \sin(122° - \theta_H) + \vec{P}_2 \cos(122° - \theta_H), \tag{4.8}$$

which forms an ellipse-shaped trace with $\vec{P}_1$ and $\vec{P}_2$ as the major and minor axes, respectively. From the $|P|$-profile, we deduced $|\vec{P}_1| \sim 31$ and $|\vec{P}_2| \sim 2\,\mu\text{C/m}^2$.

In Fig. 4.12d and e, the P-behavior expected from Eq. (4.8) is plotted as dashed lines. While the calculated P-profile roughly agrees with the observed one, small gap still exists between them. This deviation reverses its sign at $\theta_H \sim 100°$, where H becomes parallel to the magnetic easy-plane. Correspondingly, ϵ also shows small anomaly at $\theta_H \sim 100°$ (Fig. 4.12c). These behaviors can be well explained by

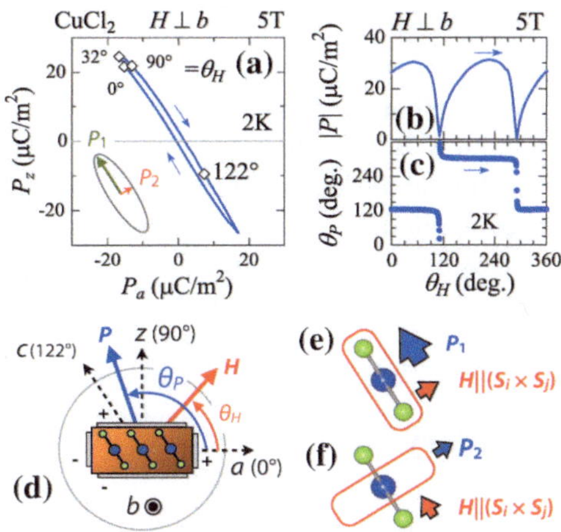

Fig. 4.13 **a** Trace of P_a and P_z under H rotating around the b-axis. **b**, **c** Magnitude and direction of P as a function of H-angle. *Arrows* indicate the direction of H-rotation. The data are taken from Fig. 4.12d and e, and the setup for measurements is shown in (**d**). θ_P (θ_H) is defined as the angle between P- (H-) direction and the a-axis. **e**, **f** The expected relationship between P, H and spin-spiral plane (depicted as *rounded square*)

assuming that the spin-spiral plane is tilted from the originally expected $\vec{H} \parallel (\vec{S}_i \times \vec{S}_j)$ position toward the magnetic easy-plane. A similar effect of magnetic-anisotropy drag on P has also been observed in the H-rotating experiment on $Eu_{1-x}Y_xMnO_3$ [33].

Thus, we conclude that the IDM scheme can well reproduce the observed FE and ME natures, even for $CuCl_2$ with $S = 1/2$ quantum spin chains. Note that $P \parallel c$ relationship observed at 0 T can be justified even with a slight revision of the originally suggested bc-spiral spin structure, since deduced ratio $|\vec{P}_1|/|\vec{P}_2| \sim 15$ is quite large. Interestingly, when H and spin-spiral plane is rotated counter-clockwise, P is found to rotate clockwise (Fig. 4.13a). This is in contrast with the case for $Eu_{1-x}Y_xMnO_3$ [33], where both H and P rotate in the same direction. The observed manner of P-rotation and large $|\vec{P}_1|/|\vec{P}_2|$ ratio are in accord with the recent DFT calculation for edge-shared CuO_2 chain compounds [34], and these features would reflect the anisotropy and sign of coupling coefficient A in Eq. (1.10).

Finally, we compare the present magnetoelectric response with that reported for $LiCuVO_4$. In $CuCl_2$, it seems that applied magnetic field above 4 T can reorient the spin-spiral plane along any direction. $H \parallel q$ induces the proper screw spin state with $P = 0$, and H orthogonal both to the original spin spiral-axis and q-vector induces the 90°-rotation of spin-spiral plane and P-direction around the q-vector, while H parallel to the original spin-spiral axis leaves both spin-spiral plane and P-direction unchanged. These behaviors perfectly agree with those reported for $LiCuVO_4$ by Schrettle et al. (Fig. 4.5) [10]. This implies that the presently observed magnetoelectric response can be commonly observed in helimagnets with similar edge-shared chain structure, with some exceptional case such as $LiCu_2O_2$ probably characterized by more complex magnetic order rather than simple cycloid. While Yasui et al. has reported the conflicting results for $LiCuVO_4$ (absence of $P \parallel c$ in the spin-flopped state under $H \parallel a$) [12], this seems to originate from the problem of measurement accuracy or crystallographic imperfection of their single crystal.

Conclusion

In this section, we have experimentally revealed magnetically-driven ferroelectricity in an $S = 1/2$ chain helimagnet $CuCl_2$, which is expected to be free from the crystallographic defects reported in Li-based compounds. Observed P-behaviors under applied H can be reproduced well within the framework of the inverse Dzyaloshinskii-Moriya model, suggesting the robustness of this ME coupling mechanism even under the effect of strong quantum fluctuation.

Summary

In this chapter, we investigated the magnetoelectric response of $S = 1/2$ chain halimagnets, and testified the validity of the inverse D-M mechanism under the strong quantum fluctuation. We briefly summarize the results obtained in each section.

$LiCu_2O_2$

By employing the polarized neutron diffraction technique, we established the existence of helimagnetic spin component within the *bc*-plane. This means that the previously reported simple *ab*-cycloid spin structure is incorrect. We also demonstrated that the reversal of P-direction always leads to the reversal of vector spin chirality ($\mathbf{S}_i \times \mathbf{S}_j$). These results resolves the reported contradiction between spin structure and observed P-direction, and proves that the inverse D-M mechanism is still effective to describe the magnetoelectric coupling in this compound.

$CuCl_2$

We investigated the magnetoelectric response of $CuCl_2$, and discovered the emergence of ferroelectricity in the cycloidal magnetic ground state. The application of $H \parallel q$ induces the disappearance of ferroelectricity, while rotation of H around the q-vector leads to the rotation of P-direction. Such complex magnetoelectric response can be well reproduced by the inverse D-M model. The observed magnetoelectric response for $CuCl_2$ is quite similar to that reported for $LiCuVO_4$ by Schrettle et al., which suggests that the conflicting results reported for $LiCuVO_4$ by Yasui et al. is probably caused by the crystallographic imperfection in their specimen.

Based on the above results, we concluded that the magnetoelectric response in $S = 1/2$ quantum chain helimagnet can be explained well within the framework of the inverse D-M model.

References

1. Furukawa S, Sato M, Onoda S (2010) Chiral order and electromagnetic dynamics in one-dimensional multiferroic cuprates. Phys Rev Lett 105:257205
2. Park S, Choi YJ, Zhang CL, Cheong S-W (2007) Ferroelectricity in an $S = 1/2$ chain cuprate. Phys Rev Lett 98:057601
3. Masuda T, Zheludev A, Bush A, Markina M, Vasiliev A (2004) Competition between helimagnetism and commensurate quantum spin correlations in $LiCu_2O_2$. Phys Rev Lett 92:177201
4. Gippius AA, Morozova EN, Moskvin AS, Zalessky AV, Bush AA, Baenitz M, Rosner H, Drechsler S-L (2004) NMR and local-density-approximation evidence for spiral magnetic order in the chain cuprate $LiCu_2O_2$. Phys Rev B 70:020406(R)
5. Masuda T, Zheludev A, Roessli B, Bush A, Markina M, Vasiliev A (2005) Spin waves and magnetic interactions in $LiCu_2O_2$. Phys Rev B 72:014405
6. Rusydi A, Mahns I, Müller S, Rübhausen M, Park S, Choi YJ, Zhang CL, Cheong S-W, Smadici S, Abbamonte P, Zimmermann M v, Sawatzky GA (2008) Multiferroicity in the spin-1/2 quantum matter of $LiCu_2O_2$. Appl Phys Lett 92:262506
7. Kobayashi Y, Sato K, Yasui Y, Moyoshi T, Sato M, Kakurai K (2009) Studies of multiferroic system of $LiCu_2O_2$: II. Magnetic structures of two ordered phases with incommensurate modulations. J Phys Soc Jpn 78:084721
8. Capogna L, Mayr M, Horsch P, Raichle M, Kremer RK, Sofin M, Maljuk A, Jansen M, Keimer B (2005) Helicoidal magnetic order in the spin-chain compound $NaCu_2O_2$. Phys Rev B 71:140402(R)
9. Gibson BJ, Kremer RK, Prokofiev AV, Assmus W, McIntyre GJ (2004) Incommensurate antiferromagnetic order in the $S = 1/2$ quantum chain compound $LiCuVO_4$. Phys B 350:e253
10. Schrettle F, Krohns S, Lunkenheimer P, Hemberger J, Büttgen N, Krug von Nidda H-A, Prokofiev AV, Loidl A (2008) Switching the ferroelectric polarization in the $S = 1/2$ chain cuprate $LiCuVO_4$ by external magnetic fields. Phys Rev B 77:144101
11. Naito Y, Sato K, Yasui Y, Kobayashi Y, Kobayashi Y, Sato M (2007) Ferroelectric transition induced by the incommensurate magnetic ordering in $LiCuVO_4$. J Phys Soc Jpn 76:023708
12. Yasui Y, Naito Y, Sato K, Moyoshi T, Sato M, Kakurai K (2008) Relationship between magnetic structure and ferroelectricity of $LiVCuO_4$. J Phys Soc Jpn 77:023712
13. Moskvin AS, Panov YD, Drechsler S-L (2009) Nonrelativistic multiferrocity in the nonstoichiometric spin-1/2 spiral-chain cuprate $LiCu_2O_2$. Phys Rev B 79:104112
14. Moskvin AS, Drechsler SL (2008) Microscopic mechanisms of spin-dependent electric polarization in $3d$ oxides. Phys Rev B 78:024102
15. Moskvin AS, Drechsler SL (2008) Multiferroicity due to nonstoichiometry in the chain cuprate $LiVCuO_4$. EPL 81:57004
16. Fang C, Datta T, Hu J (2009) Magnetoelectric coupling in the multiferroic compound $LiCu_2O_2$. Phys Rev B 79:014107
17. Yamasaki Y, Sagayama H, Goto T, Matsuura M, Hirota K, Arima T, Tokura Y (2007) Electric control of spin helicity in a magnetic ferroelectric. Phys Rev Lett 98:147204
18. Blume M (1963) Polarization effects in the magnetic elastic scattering of slow neutrons. Phys Rev 130:1670
19. Zvyagin S, Cao G, Xin Y, McCall S, Caldwell T, Moulton W, Brunel L-C, Angerhofer A, Crow JE (2002) Dimer liquid state in the quantum antiferromagnet compound $LiCu_2O_2$. Phys Rev B 66:064424
20. Onoda S, Nagaosa N (2007) Chiral spin pairing in helical magnets. Phys Rev Lett 99:027206
21. Moon RM, Riste T, Koehler WC (1969) Polarization analysis of thermal-neutron scattering. Phys Rev 181:920
22. Jia C, Onoda S, Nagaosa N, Han JH (2007) Microscopic theory of spin-polarization coupling in multiferroic transition metal oxides. Phys Rev B 76:144424
23. Capogna L, Reehuis M, Maljuk A, Kremer RK, Ouladdiaf B, Jansen M, Keimer B (2010) Magnetic structure of the edge-sharing copper oxide chain compound $NaCu_2O_2$. Phys Rev B 82:014407

24. Leininger Ph, Rahlenbeck M, Raichle M, Bohnenbuck B, Maljuk A, Lin CT, Keimer B, Weschke E, Schierle E, Seki S, Tokura Y, Freeland JW (2010) Electronic structure, magnetic, and dielectric properties of the edge-sharing copper oxide chain compound $NaCu_2O_2$. Phys Rev B 81:085111
25. Katsura H, Onoda S, Han JH, Nagaosa N (2008) Electronic structure, magnetic, and dielectric properties of the edge-sharing copper oxide chain compound $NaCu_2O_2$. Phys Rev Lett 101:187207
26. Furukawa S, Sato M, Saiga Y, Onoda S (2008) Quantum fluctuations of chirality in one-dimensional spin-1/2 multiferroics: gapless dielectric response from phasons and chiral solitons. J Phys Soc Jpn 77:123712
27. Banks MG, Kremer RK, Hoch C, Simon A, Ouladdiaf B, Broto J-M, Rakato H, Lee C, Whangbo M-H (2009) Magnetic ordering in the frustrated heisenberg chain system cupric chloride $CuCl_2$. Phys Rev B 80:024404
28. Wells AF (1947) The crystal structure of anhydrous cupric chloride, and the stereochemistry of the cupric atom. J Chem Soc 1670
29. Stout JW, Chisholm RC (1962) Heat capacity and entropy of $CuCl_2$ and $CrCl_2$ from 11 to 300 K. Magnetic ordering in linear chain crystals. J Chem Phys 36:979
30. Schmitt M, Janson O, Schmidt M, Hoffmann S, Schnelle W, Drechsler S-L, Rosner H (2009) Crystal-water-induced switching of magnetically active orbitals in $CuCl_2$. Phys Rev B 79:245119
31. Billerey D, Terrier C (1978) Magnetic behaviour of single crystal anhydrous cupric chloride in the low temperature ordered phase. Phys Lett A 68:278
32. Nagamiya T (1968) Helical spin ordering—1 theory of helical spin configurations. Solid State Phys 20:305
33. Murakawa H, Onose Y, Kagawa F, Ishiwata S, Kaneko Y, Tokura Y (2008) Rotation of an electric polarization vector by rotating magnetic field in cycloidal magnet $Eu_{0.55}Y_{0.45}MnO_3$. Phys Rev Lett 101:197207
34. Xiang HJ, Whangbo M-H (2007) Density-functional characterization of the multiferroicity in spin spiral chain cuprates. Phys Rev Lett 99:257203

Chapter 5
Summary

Conclusion

In this thesis, we investigated the magnetoelectric response in two typical forms of frustrated spin systems; triangular lattice antiferromagnets (Chap. 3) and $S = 1/2$ chain helimagnets (Chap. 4). The brief summaries for individual experiments are given at the end of each chapter. In the following, we provide the more general conclusions for each system.

Magnetoelectric Response in Triangular Lattice Antiferromagnets

- We discovered that various types of proper screw spin oder and 120°-spin order on stacked triangular lattices can induce ferroelectricity along the direction perpendicular to the spin-spiral plane, as demonstrated for $CuFe_{1-x}Ga_xO_2$, $ACrO_2$ and MnI_2. The spin structure and induced P-direction for each compound is summarized in the left side of Fig. 5.1. While this behavior can be justified from the viewpoint of the symmetry, the microscopic origin of observed magnetoelectric coupling is not trivial. Since both exchange striction and inverse D-M model denies the appearance of electric polarization for these spin textures (except for the case of MnI_2 in the low-H region), our results suggest that some other magnetoelectric coupling mechanism originating from spin-orbit coupling is active on the trigonal lattice. The observed magnetoelectric coupling can be partly explained by the recent model suggested by Arima [1] and Jia et al. [2] (i.e. modulation of hybridization strength between metal $3d$-state and oxygen $2p$-state via spin-orbit interaction), while further theoretical investigation would be indispensable for the thorough understanding.
- The symmetry of triangular lattice often hosts several equivalent multiferroic domains with different P-directions, and we found that field-induced rearrange-

Springer Theses, DOI: 10.1007/978-4-431-54091-5_5,

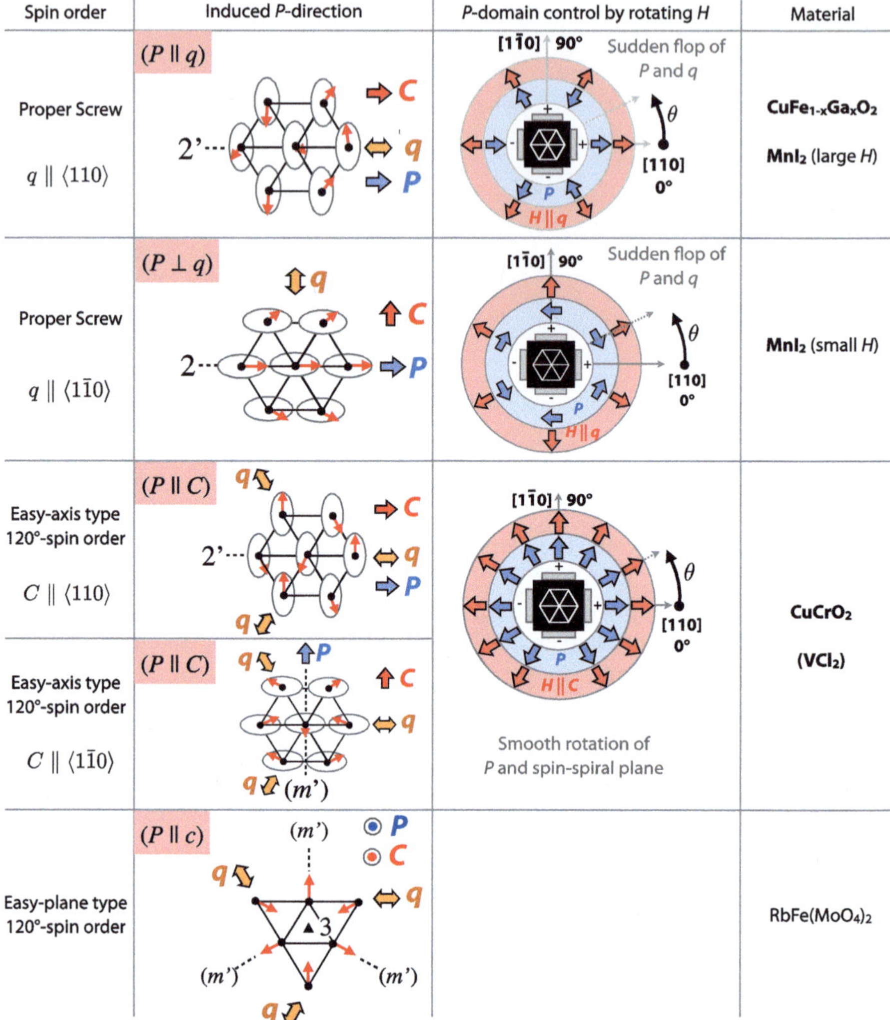

Fig. 5.1 The relationship between various types of spiral spin orders and induced P-direction on the stacked triangular lattices. Here, we define the vector spin chirality $\mathbf{C} = \mathbf{S}_i \times \mathbf{S}_j$, which is perpendicular to the spin-spiral plane. $\mathbf{q}$ denotes the in-plane component of magnetic modulation vector. The behaviors of P-vector under magnetic field rotating around the c-axis are also indicated. The materials highlighted by bold font are investigated in this thesis

ment of domain distribution can lead to unique magnetoelectric response. The observed directional change of P-vector under magnetic field rotating around the c-axis is summarized in the right side of Fig. 5.1. Here, $CuFe_{1-x}Ga_xO_2$ and MnI_2 with proper screw spin order shows discontinuous flop of P-vector (and q-vector), while $CuCrO_2$ with easy-axis type 120°-spin order shows smooth rotation of

P-vector (and spin-spiral plane). Such a difference probably reflects the strength of in-plane magnetic anisotropy and/or degree of freedom of q-vector. Notably, the chirality of spin-spiral is found to be always preserved upon the q-flop transition in $CuFe_{1-x}Ga_xO_2$ and MnI_2, although generally magnetic field cannot lift the degeneracy of two possible uniform spin-chiral state. This suggests that the choice of odd chirality upon the q-flop probably reflects the relative stability of the specific form of multiferroic domain wall.
- We also investigated the dynamical aspects of triangular lattice antiferromagnets. By employing the terahertz time-domain spectroscopy and detailed polarization analysis, we discovered an unique electromagnon (electric-field-active magnon) excitation in the paraelectric $\uparrow\uparrow\downarrow\downarrow$ collinear magnetic phase of $CuFe_{1-x}Ga_xO_2$. This electromagnon mode is found to vanish in the ferroelectric helimagnetic phase. These facts prove that neither ferroelectricity nor noncollinear magnetism is a necessary condition for the appearance of electromagnon excitation. While previous reports of electromagnon excitations are limited to ferroelectric helimagnets and are believed to arise from the exchange striction mechanism, the present anti-correlation with noncollinear magnetism excludes the exchange-striction mechanism as the origin of dynamical magnetoelectric coupling in $CuFe_{1-x}Ga_xO_2$. This implies the first experimental observation of spin-orbit coupling mediated electromagnon in the present compound, which may share the same magnetoelectric coupling mechanism with the magnetically-induced ferroelectricity in the static regime. Similar electromagnon excitation may possibly be observed for a wide variety of other paraelectric collinear magnets.

Magnetoelectric Response in $S = 1/2$ *Chain Helimagnets*

- While the one-dimensional chain helimagnets are seemingly the ideal model compounds to testify the prediction of the inverse D-M mechanism, the early two examples of such chain helimagnets with $S = 1/2$ ($LiCu_2O_2$ and $LiCuVO_4$) are reported to show magnetoelectric response inconsistent with the prediction of inverse D-M model. To testify the validity of inverse D-M model in the present system, we first performed the polarized neutron diffraction experiment on $LiCu_2O_2$. We established the existence of helimagnetic spin component within the bc-plane, and prove that the previously reported simple ab-cycloid spin structure [3] is incorrect. We also demonstrated that the reversal of P-direction always leads to the reversal of vector spin chirality ($\mathbf{S}_i \times \mathbf{S}_j$). These results resolve the reported contradiction between spin structure and observed P-direction, and prove that the inverse D-M mechanism is still effective to describe the magnetoelectric coupling in this compound. We also performed the dielectric measurements for another $S = 1/2$ chain helimagnet $CuCl_2$, and discovered magnetoelectric response quite similar to that reported for $LiCuVO_4$ by Schrettle et al. [4]. Observed behaviors are reasonably explained by the inverse D-M model, which suggests that the conflicting results reported for $LiCuVO_4$ by Yasui et al. [5] is probably caused by the crystallographic imperfection in their specimen. Based on the above results, we concluded that the magnetoelectric response in $S = 1/2$ chain helimagnets can be explained well within the framework of the inverse D-M model, even under the strong quantum fluctuation inherent in the present system.

Our present results largely expand the candidate materials which can host unique magnetoelectric response (such as non-volatile multiple-valued switching of P-direction by magnetic field or ultrafast control of magnetism by electric field), and also provide some additional guideline for the material design of such multiferroics compounds. We believe that these discoveries would contribute to the future application of multiferroics, such as H-controled FeRAM or novel optoelectronic device.

Perspective for the Future

In addition to the several problems suggested in the last section, we also present some relevant but unresolved topics for the future study.

Topological Defects in Magnetically-Induced Ferroelectrics

As discussed in "Magnetic Digital Flop of Ferroelectric Domain" and "MnI_2 with Proper Screw Spin Order" , the nature of multiferroic domain wall often plays an crucial role to determine the P-behavior under applied magnetic field. Since the properties of domain wall has been left almost unexplored even for the conventional helimagnets, further theoretical analysis as well as the direct observation of domain wall structure is highly demanded. Note that the form of defects in the ordered state depends on the underlying interactions and lattice geometry; 120° spin order on triangular lattice is suggested to host characteristic topological defects called Z_2-vortex [6, 7].

These defects are generally characterized by the additional local symmetry reduction, and are expected to host some unique properties that is not expected in the homogeneous ground state. For example, some specific form of domain wall in ferroelectric antiferromagnet $BiFeO_3$ is reported to host conductive nature, despite the insulating property in bulk phase [8]. Some other theory has also predicted the local enhancement of magnetization at the ferroelectric domain wall region [9]. The strong magnetoelectric coupling in magnetically-induced ferroelectrics should provide interesting features to various type of topological defects, and further investigation of their properties would be very promising.

Role of Anion for the Magnetically-Induced Ferroelectricity

As discussed in "MX_2-type Halides with CdI_2 Structure", MX_2 is the first example of non-chalcogen based spiral-spin induced multiferroics, while so far the study of ferroelectric helimagnets is almost limited to oxides. At this stage, the observed P-values in helimagnetic halides ($\sim$30 μC/m^2 for $CuCl_2$ and $\sim$80 μC/m^2 for MnI_2) are comparable with that observed for other helimagnetic oxides ($2,000 \sim 5$ μC/m^2). According to Jia et al. the magnitude of induced P is enhanced with stronger metal-ligand hybridization and larger spin-orbit coupling [10]. In this context, the choice of larger anion may enhance the induced P-value, since larger anion with lower electronegativity leads to smaller charge transfer gap and larger spin-orbit coupling. While the magnitude of P also depends on the lattice geometry and magnetic structure, the series of VCl_2, VBr_2, and VI_2 share the same CdI_2-type structure and easy-axis type 120°-spin order [11–13]. Systematic investigation of these compounds would offer a valuable information on how the choice of anion affects the magnitude of induced P. This may also contribute to the establishment of the general rule to obtain the larger magnetoelectric response.

References

1. Arima T (2007) Ferroelectricity induced by proper-screw type magnetic order. J Phys Soc Jpn 76:073702
2. Jia C, Onoda S, Nagaosa N, Han JH (2006) Bond electronic polarization induced by spin. Phys Rev B 74:224444
3. Masuda T, Zheludev A, Bush A, Markina M, Vasiliev A (2004) Competition between helimagnetism and commensurate quantum spin correlations in $LiCu_2O_2$. Phys Rev Lett 92:177201
4. Schrettle F, Krohns S, Lunkenheimer P, Hemberger J, Büttgen N, Krug von Nidda H-A, Prokofiev AV, Loidl A (2008) Switching the ferroelectric polarization in the $S = 1/2$ chain cuprate $LiCuVO_4$ by external magnetic fields. Phys Rev B 77:144101
5. Yasui Y, Naito Y, Sato K, Moyoshi T, Sato M, Kakurai K (2008) Relationship between magnetic structure and ferroelectricity of $LiVCuO_4$. J Phys Soc Jpn 77:023712
6. Kawamura H, Miyashita S (1984) Phase transition of the two-dimensional heisenberg antiferromagnet on the triangular lattice. J Phys Soc Jpn 53:9
7. Kawamura H, Miyashita S (1984) Phase transition of the two-dimensional heisenberg antiferromagnet on the triangular lattice. J Phys Soc Jpn 53:4138
8. Seidel J, Martin LW, He Q, Zhan Q, Chu Y-H, Rother A, Hawkridge ME, Maksymovych P, Yu P, Gajek M, Balke N, Kalinin SV, Gemming S, Wang F, Catalan G, Scott JF, Spaldin NA, Prenstein J, Ramesh R (2009) Conduction at domain walls in oxide multiferroics. Nat Mater 8:229
9. Daraktchiev M, Catalan G, Scott JF (2008) Landau theory of ferroelectric domain walls in magnetoelectrics. Ferroelectrics 375:122
10. Jia C, Onoda S, Nagaosa N, Han JH (2007) Microscopic theory of spin-polarization coupling in multiferroic transition metal oxides. Phys Rev B 76:144424
11. Kadowaki H, Ubukoshi K, Hirakawa K, Martinez JL, Shirane G (1987) Experimental study of new type phase transition in triangular lattice antiferromagnet VCl_2. J Phys Soc Jpn 56:4027
12. Kadowaki H, Ubukoshi K, Hirakawa K (1985) Neutron scattering study of the triangular-lattice antiferromagnet VBr_2. J Phys Soc Jpn 54:363
13. Hirakawa K, Kadowaki H, Ubukoshi K (1983) Study of frustration effects in two-dimensional triangular lattice antiferromagnets : neutron powder diffraction study of VX_2, X = Cl, Br and I. J Phys Soc Jpn 52:1814

Publication List

Original Paper

1. "Magnetic-field induced competition of two multiferroic orders in a triangular-latticehelimagnet MnI_2"
   T. Kurumaji, S. Seki, S. Ishiwata, H. Murakawa, Y. Tokunaga, Y. Kaneko, and Y. Tokura
   Phys. Rev. Lett. **106**, 167206 (2011)
2. "Electromagnons in the spin collinear state of a triangular lattice antiferromagnet"
   S. Seki, N. Kida, S. Kumakura, R. Shimano, and Y. Tokura
   Phys. Rev. Lett. **105**, 097207 (2010)
3. "Cupric chloride $CuCl_2$ as an $S = 1/2$ chain multiferroic"
   S. Seki, T. Kurumaji, S. Ishiwata, H. Matsui, H. Murakawa, Y. Tokunaga, Y.Kaneko, T. Hasegawa, and Y. Tokura
   Phys. Rev. B, **82**, 064424 (2010).
4. "Electronic structure, magnetic, and dielectric properties of the edge-sharingcopper oxide chain compound NaCu2O2"
   Ph. Leininger, M. Rahlenbeck, M. Raichle, B. Bohnenbuck, A. Maljuk, C. T.Lin, B. Keimer, E.Weschke, E. Schierle, S. Seki, Y. Tokura, and J. W. Freeland
   Phys. Rev. B **81**, 085111 (2010).
5. "Magnetic digital flop of ferroelectric domain with fixed spin chirality in atriangular lattice helimagnet"
   S. Seki, H. Murakawa, Y. Onose, and Y. Tokura
   Phys. Rev. Lett. 103, 237601 (2009).
6. "Spin-driven ferroelectricity in triangular lattice antiferromagnets $ACrO_2$ (A= Cu, Ag, Li, or Na) "
   S. Seki, Y. Onose, and Y. Tokura
   Phys. Rev. Lett. **101**, 067204 (2008).
7. "Correlation between spin helicity and electric polarization vector in quantumspinchain magnet $LiCu_2O_2$"

Springer Theses, DOI: 10.1007/978-4-431-54091-5,

S. Seki, Y. Yamasaki, M. Soda, M. Matsuura, K. Hirota, Y. Tokura
Phys. Rev. Lett. **100**, 127201 (2008).

8. "Impurity-doping-induced ferroelectricity in the frustrated antiferromagnet $CuFeO_2$"
S. Seki, Y. Yamasaki, Y. Shiomi, S. Iguchi, Y. Onose, and Y. Tokura
Phys. Rev. B **75**, 100403(R) (2007).

Review

1. "Multiferroics with spiral spin orders"
Y. Tokura and S. Seki
Adv. Mater. **22**, 1554 (2010).

MIX
Papier aus verantwortungsvollen Quellen
Paper from responsible sources
FSC® C105338

If you have any concerns about our products,
you can contact us on
ProductSafety@springernature.com

In case Publisher is established outside the EU,
the EU authorized representative is:
Springer Nature Customer Service Center GmbH
Europaplatz 3, 69115 Heidelberg, Germany

Printed by Libri Plureos GmbH
in Hamburg, Germany